ORTHOGONAL METHODS
FOR ARRAY SYNTHESIS

ORTHOGONAL METHODS FOR ARRAY SYNTHESIS

Theory and the ORAMA Computer Tool

John N. Sahalos

Radiocommunications Laboratory
University of Thessaloniki, Greece

John Wiley & Sons, Ltd

Other Wiley Editorial Offices

John Wiley & Sons Inc., 111 River Street, Hoboken, NJ 07030, USA

Jossey-Bass, 989 Market Street, San Francisco, CA 94103-1741, USA

Wiley-VCH Verlag GmbH, Boschstr. 12, D-69469 Weinheim, Germany

John Wiley & Sons Australia Ltd, 42 McDougall Street, Milton, Queensland 4064, Australia

John Wiley & Sons (Asia) Pte Ltd, 2 Clementi Loop #02-01, Jin Xing Distripark, Singapore 129809

John Wiley & Sons Canada Ltd, 22 Worcester Road, Etobicoke, Ontario, Canada M9W 1L1

Wiley also publishes its books in a variety of electronic formats. Some content that appears in print may not be available in electronic books.

British Library Cataloguing in Publication Data

A catalogue record for this book is available from the British Library

ISBN-13 978-0-470-01741-8 (HB)
ISBN-10 0-470-01741-4 (HB)

Typeset in 10/12pt Times by Laserwords Private Limited, Chennai, India.
Printed and bound in Great Britain by Antony Rowe Ltd, Chippenham, Wiltshire.
This book is printed on acid-free paper responsibly manufactured from sustainable forestry in which at least two trees are planted for each one used for paper production.

To my wife Angela for her love, patience and devotion
Στη γυναίκα μου Αγγελική για την αγάπη, υπομονή και αφοσίωσή της

Contents

Preface

Orthogonal methods for array synthesis are designed to meet the needs of antenna engineers and graduate students of Electrical and Computer Engineering and Physics Departments. The objective of the present book is to describe a relatively simple approach to the design of antenna arrays.

Essentially, the book represents the theory and applications of the Orthogonal Methods (OM), including the corresponding details of over two-decades of comprehensive research activity at the Radiocommunications Laboratory at the University of Thessaloniki, Greece. The use of plural in the term 'Orthogonal Methods', instead of 'Method', was considered to be preferable. That was justified by the fact that the orthogonalization of the vector space in array synthesis contains different procedures for different purposes. A considerable attempt to make the book as complete and self-contained as possible has been made.

The five chapters of the book address the following topics:

Chapter 1 is an introductory chapter that contains the basics and theory of antennas and antenna arrays. Chapter 2 gives an overview of the most common antenna arrays. Linear arrays with uniform excitations as well as Chebyshev and other non-uniform excitations are presented. Furthermore, planar, rectangular and circular as well as 3-D and conformal arrays are also outlined. Chapter 3 contains most of the well-known array design techniques. The pattern synthesis of several uniform and non-uniform arrays is explained. Array design procedures by sampling or by root matching of Taylor and Bayliss line sources are analyzed. Moreover, matrix methods, simplex and gradient methods, simulated annealing and genetic algorithms are discussed for the design of antenna array systems. Antenna arrays combined with signal processing, known as 'smart antennas', are also given. Chapter 4 describes the Orthogonal Methods. After an introduction to the essential works of Unz and Uzkov, the classical orthogonal method with and without constraints for linear, planar and 3-D arrays is presented. Moreover, the orthogonal method with the method of moments and the orthogonal compensation method are given. The conformal orthogonal method is also presented. Finally, the orthogonal perturbation method for the geometry synthesis is described. The chapter details extensively the applications of the OM to the design of antenna arrays with a wide list of array patterns.

In the last chapter, the ORAMA computer tool, with the instructions on how to use the material in the accompanying CD, is explained. ORAMA, in the given version, contains the programs of the classical Orthogonal Method for synthesis of linear arrays. The design cases cover a sufficient number of needs of antenna engineers and graduate students.

I would like to acknowledge the invaluable contribution of my two colleagues and ex-students George Miaris and Sotiris Goudos. Both of them have developed most of the computer programs that are included in ORAMA. George Miaris has also helped in co-authoring Chapter 5, as well as in proofreading and revising the manuscript until the last day of submission. Also, many graduate students from the University of Thessaloniki have been involved in orthogonal methods. Among them the contribution of Christos Iakovidis should be recognized. I would also like to mention the work of my graduate students Christos Kalialakis, Apostolos Georgiadis, Emmanouil Volidis, Christos Tsironas, Michael Polychronidis and Vasilis Varsamis. Special thanks are addressed to my colleagues Professors Elias Vafiadis, Katerina Siakavara, Theodoros Samaras, Dr. Dimitrios Babas, Dr. Theodoros Kaifas and Dr. Zaharias Zaharis for their long-term support in the RCLab. I would also like to express my gratitude to Lena Gialabouki who has read and edited the manuscript. Moreover, the author wishes to acknowledge the constructive criticism, encouragement and friendship of Prof. Constantine A. Balanis.

Last but not least, I would like to thank my wife Angela and my daughters Eleftheria, Stavroula and Sofia for their love, patience, support and understanding for the many hours of neglect during the writing of this book.

John N. Sahalos
University of Thessaloniki
Greece.

1

Antennas and Antenna Arrays

1.1 Introduction

Antennas have become ubiquitous devices and occupy a salient position in wireless systems. Radio and TV as well as satellite and new generation mobile communications have experienced the largest growth among industry systems. The global wireless market continues to grow at breakneck speed and the strongest economic and social impact nowadays comes from cellular telephony, personal communications and satellite navigation systems. All of the above systems have served as motivation for engineers to incorporate elegant antennas into handy and portable systems.

Many textbooks provide in-depth resources on antennas. Especially on antenna arrays, there are digests, studies and books containing extensive data and techniques. In the references given herewith [1–26], there are some of the best-known and most highly recommended books.

A device able to receive or transmit electromagnetic energy is called an 'antenna'. As seen in [6], the antenna plays the role of a *transitional structure between free space and a guiding device*. An antenna consists of one or more elements. A single-element antenna is usually not enough to achieve technical needs. That happens because its performance is limited. A set of discrete elements, which constitute an antenna array, offers the solution to the transmission and/or reception of electromagnetic energy. The geometry and the type of elements characterize an antenna array. For simplicity, implementation and fabrication reasons, the elements are chosen in such a way so as to be identical and parallel. For the same reasons, uniformly spaced linear arrays are mostly encountered in practice.

In the following paragraphs, the properties of various antenna arrays will be presented.

1.2 Antenna Array Factor

The radiation characteristics of antennas have mostly to do with the far field (*Fraunhofer*) region. In this region, the field expression is a multiplication of two parts. One part contains the distance r dependence of the observation point (receiver location) and the other contains its spherical coordinate angles θ and ϕ dependence. The angular distribution

Orthogonal Methods for Array Synthesis: Theory and the ORAMA Computer Tool   John N. Sahalos
© 2006 John Wiley & Sons, Ltd

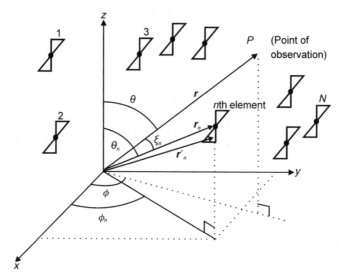

Figure 1.1   Geometry of an antenna array with N identical and parallel elements

of the field is independent of the distance r. For a typical antenna element (see Fig. 1.1), the far electric field is

$$E_n(r) \cong -j\omega\mu \frac{e^{-j\beta r}}{4\pi r} f_n(\theta, \phi) \tag{1-1}$$

The angular-dependent vector $f_n(\theta, \phi)$ gives the directional characteristics of the nth element electric field [11]:

$$f_n(\theta, \phi) = (\hat{\theta}\hat{\theta} + \hat{\phi}\hat{\phi}) \cdot \int_{\text{element}} J_n(r'_n) e^{j\beta\hat{r}\cdot(r_n - r'_n)} \, dv' \tag{1-2}$$

where

$J_n(r'_n)$ = electric current density of the nth element

r'_n = distance of a source point from the origin

r = distance of the observation point from the origin

$\beta = \dfrac{2\pi}{\lambda}$ the free space wave number

ω = the angular frequency and

μ = the magnetic permeability of the space

The total electric field of an N element antenna array is

$$E(r) = \sum_{n=1}^{N} E_n(r) \tag{1-3}$$

Moreover, the total magnetic field is [6],

$$H(r) = \frac{1}{\eta}\hat{r} \times E(r) \tag{1-4}$$

where $\eta = \sqrt{\mu/\epsilon}$ (ϵ is the electric permeability of the space).

For identical and identically oriented elements, the current distribution of each element is approximately the same except for a constant complex multiplier. In (1-1), $f_n(\theta, \phi)$ can be expressed as

$$f_n(\theta, \phi) = I_n f(\theta, \phi) \tag{1-5}$$

$f(\theta, \phi)$ is called the 'pattern function' of the element and I_n is the complex excitation of the nth element of the array.

(1-1), (1-2) and (1-5) are combined and give

$$E(r) = -j\omega\mu \frac{e^{-j\beta r}}{4\pi r} f(\theta, \phi) \sum_{n=1}^{N} I_n e^{j\beta r_n \cos \xi_n} \tag{1-6}$$

(r_n, θ_n, ϕ_n) are the spherical coordinates of a convenient reference point of the nth element and $\cos \xi_n = \sin\theta \sin\theta_n \cos(\phi - \phi_n) + \cos\theta \cos\theta_n$.

The last term of (1-6) is expressed separately as

$$AF(\theta, \phi) = \sum_{n=1}^{N} I_n e^{j\beta r_n \cos \xi_n} \tag{1-7}$$

$AF(\theta, \phi)$ is called the 'array factor'. This factor is actually the array pattern of N isotropic point sources positioned at the reference points of the elements of the original array.

From (1-6) and (1-7), we have the following:

$$E(r) = -j\omega\mu \frac{e^{-j\beta r}}{4\pi r} f(\theta, \phi) AF(\theta, \phi) \tag{1-8}$$

Expression (1-8) states the following pattern multiplication principle: *An array consisting of identical and identically oriented elements has a pattern, which can be expressed as the product of the element pattern and the array factor.*

An antenna engineer has to make an anticipatory and compatible choice of elements according to technical requirements. Once the element pattern is derived, the design effort is mainly directed at the array factor.

1.3 Elements and Array Types

Element types of antenna arrays are delineated in the literature [1–14]. Dipoles, monopoles, loops, slots, microstrip patches and horns are the most common types of array elements. Recent studies and innovations have resulted in new types of elements. Some of them are the monolithic, the superconducting, the active and the electronically and functionally small antenna elements.

In parallel with the development of elements, antenna arrays have experienced a tremendous growth. Their list starts with the linear broadside and end-fire arrays, the planar, the circular, and the conformal, and goes up to the adaptive arrays. Moreover, flat plate slot arrays, digital beam forming, dichroic, slotted and fractal arrays are some of the recent types.

It was mentioned previously that antenna analysis and synthesis focuses mostly on the array factor. Consequently, in the following paragraphs we devote the analysis mainly to this factor.

1.4 Antenna Parameters and Indices

In many cases, it is necessary to characterize the performance of antennas by referring to specific parameters and indices. Most of these parameters and indices have been defined by the committees of the institutions in charge (IEEE, ETSI etc.) and will be presented in this paragraph.

It is well known that antennas have to do with applications of time varying fields. Sources with $e^{j\omega t}$ time variation produce fields that also vary in the same way. In the literature [6, 22], the above fields are called 'time-harmonic fields'. If a signal with certain bandwidth is present in an antenna, the Fourier transform can derive the time varying forms of the electromagnetic quantities. The procedure is analogous to that of solving electric circuit problems. In this book, time-harmonic fields with a proper choice of working frequency/ies are assumed.

1.4.1 Radiated Power

The time average power density, which is the average Poynting vector, can be written in the following form:

$$S = \frac{1}{2}\text{Re}[E \times H^*] \tag{1-9}$$

The $1/2$ in (1-9) appears because the electric and magnetic fields represent the peak values.

The radiated power P_r, through a closed surface S surrounding the antenna, can be obtained by integrating the normal component of the average Poynting vector over the entire surface. On the basis of the above definition we have

$$P_r = \oiint_s S \cdot ds = \frac{1}{2} \oiint_s \text{Re}[E \times H^*] \cdot ds \tag{1-10}$$

The surrounding surface can be of arbitrary shape, and without losing generality a sphere is used.

1.4.2 Radiation Intensity

Radiation intensity in a given direction is *the power radiated per unit solid angle in the above-mentioned direction*. It is expressed in watts per unit solid angle and is related to the far field of the antenna. In a spherical coordinate system (r, θ, ϕ), radiation intensity is defined as

$$U(\theta, \phi) = r^2 S \cdot \hat{r} \tag{1-11}$$

where r denotes the distance between the antenna and the observation point and $\hat{r}$ is the corresponding radial unit vector.

$U(\theta, \phi)$ can be expressed only by the electric or the magnetic far-zone field:

$$U(\theta, \phi) = \frac{r^2}{2\eta}|E(r, \theta, \phi)|^2 = \eta\frac{r^2}{2}|H(r, \theta, \phi)|^2 \tag{1-12}$$

where η is the intrinsic impedance of the medium. (For the free space $\eta = 120\pi$ Ohms).

The total power P_r radiated is obtained by taking the integral of the radiation intensity over all angles around the antenna. Thus, P_r is

$$P_r = \oint\!\!\!\oint U\,d\Omega = \int_0^{2\pi}\int_0^{\pi} U(\theta,\phi)\sin\theta\,d\theta\,d\phi \qquad (1\text{-}13)$$

Another parameter related to $U(\theta,\phi)$ is the average radiation intensity, defined as the radiation intensity of an isotropic source radiating the same power as that of the actual antenna. Thus,

$$U_{ave} = \frac{1}{4\pi}\oint\!\!\!\oint_{\Omega} U\,d\Omega = \frac{P_r}{4\pi} \qquad (1\text{-}14)$$

where $d\Omega$ is the element of solid angle.

It is obvious that an isotropic source (antenna) is not realizable; however, it is often used as a reference element for many antennas.

1.4.3 Directivity

The directivity of an antenna in a given direction is defined as the ratio of the radiation intensity in the above-mentioned direction to the average radiation intensity. If the direction is not specified, then it is implied in the direction of maximum radiation. By using (1-12) and (1-13), directivity D is

$$D = \frac{U}{U_{ave}} \qquad (1\text{-}15)$$

Directivity can be more conveniently expressed by

$$D = 4\pi\frac{U}{P_r} \qquad (1\text{-}16)$$

Maximum directivity is given by

$$D_{\max} = D_o = 4\pi\frac{U_{\max}}{P_r} \qquad (1\text{-}17)$$

The directivity of an isotropic source is unity. All other sources have maximum directivity greater than unity.

1.4.4 Antenna Efficiency

It is well known that only one part of the input power at the input terminals of an antenna is radiated. Various reasons, like the mismatch between the transmission line and the antenna or conduction and dielectric losses of the antenna itself, reduce the power radiated.

The total efficiency of an antenna can be expressed by

$$e_o = e_r e_c e_d \qquad (1\text{-}18)$$

where e_o is the total efficiency, e_r is the mismatch efficiency $= (1 - |\Gamma|^2)$, e_c is the conduction efficiency and e_d is the dielectric efficiency. Γ is the voltage reflection coefficient at the input of the antenna terminals expressed by

$$\Gamma = \frac{Z_{in} - Z_o}{Z_{in} + Z_o} \tag{1-19}$$

where Z_{in} is the antenna input impedance and Z_o is the characteristic impedance of the transmission line.

(1-18) is expressed in an alternative form as

$$e_o = e_{cd}e_r = e_{cd}(1 - |\Gamma|^2) \tag{1-20}$$

e_{cd} is the radiation efficiency that can be determined experimentally or, if possible, numerically.

1.4.5 Gain

The performance of an antenna can also be described by the gain. The gain is related to directivity. It is an index that takes the directional properties and the efficiency of the antenna into account. The gain $G(\theta, \phi)$ is defined by

$$G(\theta, \phi) = \frac{\text{radiation intensity for the direction}(\theta, \phi)}{\dfrac{1}{4\pi}(\text{power input to the antenna})} \tag{1-21}$$

When the direction is not specified, the gain is taken in the direction of maximum radiation. Expression (1-21) counts the losses of the antenna itself and can be written as

$$G(\theta, \phi) = e_{cd}\frac{U(\theta, \phi)}{\dfrac{1}{4\pi}P_r} = e_{cd}D \tag{1-22}$$

If one takes the mismatch losses into account, then the absolute gain G_{abs} is introduced [6]. G_{abs} is expressed by

$$G_{abs} = e_o D \tag{1-23}$$

Finally, maximum absolute gain $G_{o_{abs}}$ is related to maximum directivity D_o. It is

$$G_{o_{abs}} = e_o D_o \tag{1-24}$$

The above indices are given in terms of decibels instead of dimensionless quantities. The relation of an index X in dB and its corresponding dimensionless value is

$$X[\text{dB}] = 10 \log X[\text{dimensionless}] \tag{1-25}$$

1.4.6 Antenna Patterns

One of the main characteristics of an antenna is its radiation pattern. It presents graphically the radiation properties and can be measured by moving a probe antenna around the antenna under test at a constant distance in the far field (see Fig. 1.2a). The response, as a function of the angular coordinates, constitutes the radiation pattern. Depending on

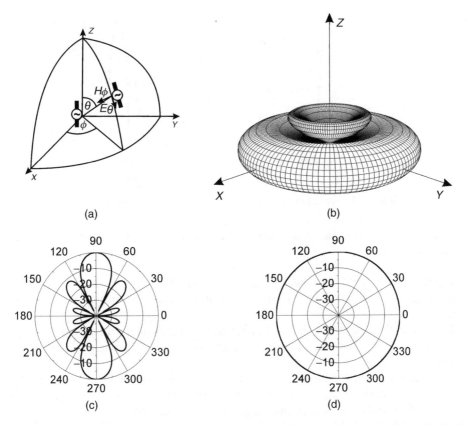

Figure 1.2 Radiation pattern of a linear array of three vertical collinear electric $\lambda/2$ dipoles in $d/\lambda = 0.75$ (a) Pattern measurement scheme, (b) Three-dimensional plot, (c) E-plane radiation pattern, (d) H-plane radiation pattern

probe type and orientation, the appropriate component of an electric or a magnetic field can be measured. If a probe is moved over the spherical surface, its terminal voltage will present the 3-D radiation pattern. A pattern taken on one plane is known as a 'plane pattern'. The pattern that contains the electric field vector is the E-plane pattern while the pattern that contains the magnetic field vector is the H-plane. The above two are referred to as the 'principal plane patterns'. As an example, Fig. 1.2b presents the 3-D radiation pattern of a uniform linear array of three collinear vertical electric dipoles with equal mutual distance of 0.75λ. Figures 1.2c and d show the E- and H-plane patterns of the array.

A plane pattern can be depicted as a polar or as a rectangular plot. The units of the patterns are either linear (Fig. 1.3a) or logarithmic (dB) (Fig. 1.3b). The lobe in the direction of maximum radiation is called the 'main lobe' or the 'main beam'. Any lobe of the pattern other than the main lobe is a 'minor lobe'. Usually, a minor lobe is also called a 'side lobe'. A side lobe in a pattern is, in general, any lobe other than that of the intended one. Since the intended lobe is usually the main lobe, it is obvious that minor

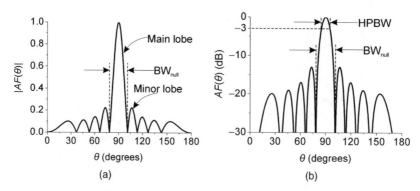

Figure 1.3   Radiation patterns in (a) linear and (b) logarithmic scale

and side lobes are the same. A measure of the characteristics of the pattern is given by certain quantities:

1. The half-power (or 3 dB power) beam width (HPBW) is the angular width between the angular points half-power (3 dB) below that of the main-beam maximum of the antenna (see Fig. 1.3b).
2. The first null beam width (BW_{null}) is defined as the angular width between the first zero crossing of either side of the main-beam maximum of the antenna (see Figs. 1.3a and b).
3. The bandwidth Δf of an antenna is defined by the frequency limits at which the maximum gain is reduced to its half value (3 dB). The fractional bandwidth is given by $\Delta f / f$. f is the mean operating frequency of the antenna.
4. The side lobe level (SLL) is the ratio of the pattern value of a side lobe peak to the corresponding value of the main lobe. Usually, SLL in an antenna is defined as the largest side lobe level for the whole pattern. A special case of the inverse SLL is the front to back ratio (F/B). This is the ratio of the pattern value in the main lobe maximum to the corresponding value in the direction of 180 degrees from it. If there is a minor lobe in the back direction, this is called a 'back lobe' (see Fig. 1.4a).

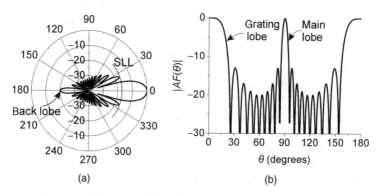

Figure 1.4   (a) Polar plot of an antenna pattern with $SLL = -17.24$ dB and $F/B = 20$ dB, (b) Rectangular plot of an antenna with grating lobes

5. The grating lobe is any maximum equal to the maximum of the main lobe of the pattern. One or more grating lobes are formed in antenna arrays when the spacing between the elements is more than λ (see Fig. 1.4b).

Depending on its radiation pattern, an antenna is called 'broadside', 'end-fire' or 'intermediate'. A broadside antenna is an antenna for which the direction of the main lobe maximum is normal to the plane containing the antenna. If the above direction is within the plane, then the antenna is an end-fire antenna. An antenna is intermediate if it is neither broadside nor end-fire. Figures 1.5a, b and c represent the pattern of the three types of antennas. The beam of the antenna in Fig. 1.5a is known as a 'fan beam'. The single lobe of the antenna in Fig. 1.5b is called a 'pencil beam'.

It is noticed that, owing to reciprocity, the radiation pattern of an antenna in the transmitting mode is identical to that in the receiving mode. The reciprocity theorem is well known from circuit analysis [27]. It states: *If a constant current (voltage) source is placed in one branch of a reciprocal network, and a voltage (current) reaction is measured in another, then interchanging the locations between the source and the branch of measurement, we have unchanged measurement results.* A network is reciprocal when it is composed of linear, bilateral elements. In antennas, the materials of the construction, the transmission line and the medium of wave propagation must be linear. Nonlinear devices (diodes, transistors) make the antenna nonreciprocal. In electromagnetics, the reciprocity

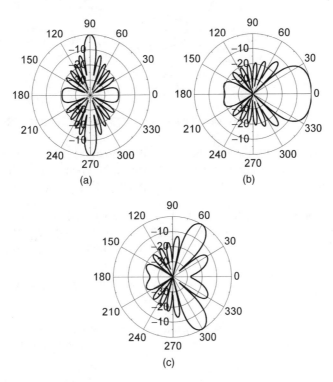

Figure 1.5 Radiation pattern of (a) A broadside array (fan beam), (b) An end-fire array (pencil beam) and (c) An intermediate array

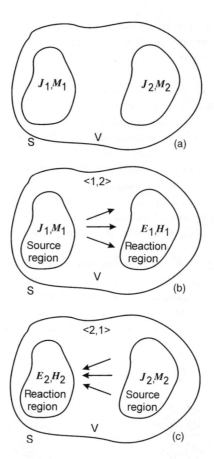

Figure 1.6   (a) Geometry of two reacted sources (b) Reaction $\langle 1, 2 \rangle$ and (c) Reaction $\langle 2, 1 \rangle$

theorem is represented with the help of Maxwell's equations and is called the 'Lorentz reciprocity theorem'. Application of the above theorem to open (unbounded) problems is of importance to antennas. Without missing the point of this book, it would be useful to give some more details.

Two pairs of sources, (J_1, M_1) and (J_2, M_2), existing in a volume V bounded by the surface S (see Fig. 1.6a), associated correspondingly with the fields (E_1, H_1) and (E_2, H_2), follow the reciprocity theorem, which, for S of infinite radius, is expressed by

$$\iiint_V (E_1 \cdot J_2 - H_1 \cdot M_2)\, dV = \iiint_V (E_2 \cdot J_1 - H_2 \cdot M_1)\, dV \qquad (1\text{-}26)$$

Each of the integrals is interpreted as a coupling between the fields of a set of sources and another set of sources, which generate another set of fields. The quantity (see Fig. 1.6b)

$$\langle 1, 2 \rangle = \iiint_V (E_1 \cdot J_2 - H_1 \cdot M_2)\, dV \qquad (1\text{-}27)$$

is called a 'reaction of the field (E_1, H_1) to the sources (J_2, M_2)' [28]. In a similar way, the quantity (see Fig. 1.6c)

$$\langle 2, 1 \rangle = \iiint\limits_{V} (E_2 \cdot J_1 - H_2 \cdot M_1)\, dV \tag{1-28}$$

is called a 'reaction of the field (E_2, H_2) to the sources (J_1, M_1)'. From (1-26)–(1-28), one can write that

$$\langle 1, 2 \rangle = \langle 2, 1 \rangle \tag{1-29}$$

In electromagnetics, the quantity to be measured is the field due to one source and the branch of the network is the volume containing the other source. From (1-29), it is obvious that reaction is unchanged by the interchange of source and measurement locations.

Besides the radiation pattern, another useful diagram in antennas is the polarization pattern. The polarization of an antenna is the polarization of the wave transmitted by the antenna. Polarization corresponds to a certain direction. If the direction is not stated, then it is assumed that it corresponds to the direction of the maximum. Polarization is characterized by the curve traced by the end point of the arrow representing the instantaneous electric field. The field is observed in the direction of propagation. Polarization is classified as 'linear', 'circular' or 'elliptical'. The polarization pattern of an antenna in a certain direction is defined as *the angular response of a test probe rotated around the axis of the given direction of the antenna under test*. In Fig. 1.7, the polarization measuring system is presented while, in Fig. 1.8, the pattern for linear, circular and elliptical polarization is given. At the radiation sphere around an antenna, polarization contains a pair of two orthogonal components. One of the components is 'co-polarization' and the other is 'cross polarization'. To accomplish the components, co-polarization is first specified at each point on the radiation sphere. After this, cross polarization is specified. For linearly polarized antennas, the orientation of the co-polar electric field in the maximum of the main beam is first specified. This maximum is directed along the polar axis of the radiation sphere. In the general case of an antenna with elliptical polarization, the major axes of the ellipse are defined. The polarization can be displayed on the surface of the Poincare sphere [29, 30]. The details of the sphere and the appropriate procedures are given in detail in the literature [31].

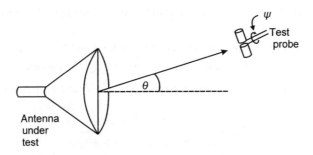

Figure 1.7   Polarization measuring system

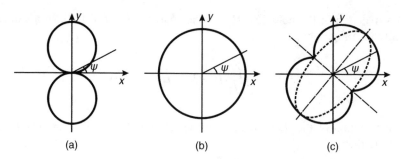

Figure 1.8 Polarization pattern (a) Linear, (b) Circular, (c) Elliptical

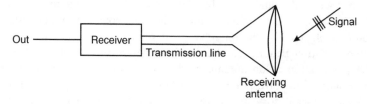

Figure 1.9 Receiving antenna with the impinged wave

1.4.7 Antenna Effective Aperture

The receiving antennas are mainly associated with their capturing characteristics. If a wave impinges on a receiving antenna (see Fig. 1.9) from a direction (θ, ϕ), its ability to capture power can be derived by a number of equivalent areas [6].

The effective aperture (area) is one of the most common quantities. It is

$$A_e = \frac{\text{Power available at the receiving antenna terminals}}{\text{Power density of incident wave}} \quad (1\text{-}30)$$

It is assumed that the wave polarization is matched to the antenna.

Under matching conditions, the effective aperture becomes maximum and is symbolized by A_{em}. The maximum effective area is related to the maximum directivity D_o of the antenna. It is

$$A_{em} = \frac{\lambda^2}{4\pi} D_o \quad (1\text{-}31)$$

1.4.8 Beam Efficiency

Beam efficiency (BE) is a dimensionless quantity of an antenna related to the main beam characteristics. BE is defined as

$$\text{BE} = \frac{\text{Power transmitted (received) within the cone angle } \theta_h}{\text{Power transmitted (received) by the antenna}} \quad (1\text{-}32)$$

If the main-beam maximum is directed along the z-axis and θ_h is the cone angle, then (see Fig. 1.10)

$$\text{BE} = \frac{\int_0^{2\pi} \int_0^{\theta_h} U(\theta, \phi) \sin \theta \, d\theta \, d\phi}{\int_0^{2\pi} \int_0^{\pi} U(\theta, \phi) \sin \theta \, d\theta \, d\phi} \quad (1\text{-}33)$$

Usually, the angle θ_h is chosen at the first null or minimum of the main beam.

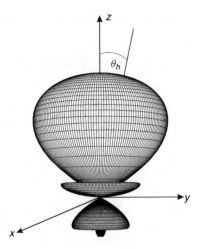

Figure 1.10 The cone angle θ_h, where a fraction of the total radiation power of an antenna exists

Besides BE, the main-beam radiation efficiency (n) is used for antenna arrays. It is a quantity useful for measuring the radiation intensity in the maximum of the main beam versus the total power of the current excitation of the elements of an array. For an array with N elements, n is expressed as

$$n = \frac{\left|\sum_{i=1}^{N} I_i\right|^2}{N \sum_{i=1}^{N} |I_i|^2} \tag{1-34}$$

In [32], it has been shown that the maximum value $n = 1$ is possible for uniform cophasal excitation.

1.4.9 Signal-to-Noise Ratio

The signal-to-noise ratio (SNR) is a dimensionless quantity related to receiving antennas, which is defined as

$$\text{SNR} = \frac{\text{received power of the desired signal}}{\dfrac{1}{4\pi}(\text{noise-plus-interference power})} \tag{1-35}$$

SNR is related to the antenna gain $G(\theta, \phi)$ and the noise distribution. For uniform noise distribution, SNR is proportional to the gain.

1.4.10 Quality Factor

Antenna performance can be evaluated in various ways. Some figures of merit that are used for antennas include quantities such as effective aperture, gain, bandwidth, efficiency, noise power and quality factor (QF). The above are interrelated and cannot be independently optimized.

The quality factor, represented by the symbol Q, is an index that measures antenna efficiency. It is defined as

$$Q = \frac{\omega \cdot \text{average stored energy}}{\text{average power dissipated}} \tag{1-36}$$

The above definition is the same one used in electric circuits. In an antenna, there are radiation, conduction, dielectric and surface wave losses. The quality factor contains all these losses and is written as [33],

$$\frac{1}{Q_t} = \frac{1}{Q_r} + \frac{1}{Q_c} + \frac{1}{Q_d} + \frac{1}{Q_{sw}} \tag{1-37}$$

Q_t is the total QF, Q_r is the QF due to radiation, Q_c is the QF due to conduction losses, Q_d is the QF due to dielectric losses and Q_{sw} is the QF due to surface waves.

For a capacitive antenna, Q_t is $1/\omega RC$, where R includes radiation and loss resistance.

When Q_t is large, the quality factor can be interpreted as the reciprocal of the frequency bandwidth of the antenna. Thus,

$$Q = \frac{f_o}{\Delta f} \tag{1-38}$$

As in circuit theory, f_o is the resonant frequency and Δf is the bandwidth for which the resistance is equal to the reactance.

In some cases, the ratio G/Q_t is used as a figure of merit. This is known as a gain - bandwidth product owing to the above-mentioned reciprocity.

For an antenna array, the so-called array Q factor is given by

$$Q = \frac{\text{sum of the excitation current magnitude squared}}{\dfrac{1}{4\pi}(\text{power input to the array})} \tag{1-39}$$

The above expression comes from the usual meaning of the definition given in the beginning of 1.4.10. (1-21), (1-34) and (1-39) are combined for an N element array and give

$$G_{O_{abs}} = NnQ \tag{1-40}$$

1.4.11 Sensitivity Factor

An important index useful for the practical implementation of antenna arrays is the sensitivity factor K. This factor is given by [34],

$$K = \frac{\text{sum of the excitation magnitude squared}}{\text{square of the magnitude of the array factor at } \xi_n = 0} \tag{1-41}$$

An ideal array is different from a real one. When an array is built, the pattern is not precisely realized to the ideal pattern. The difference comes from the currents and the location of the elements, which may differ from the desired ones.

The K factor measures the susceptibility of the pattern to random errors in the position and/or the excitation of the elements of the array.

1.5 Antenna Input Impedance

Antenna 'input impedance' (Z_{in}) is the impedance of an antenna at its terminals. Alternatively, Z_{in} is the ratio of the voltage (electric field) to the current (magnetic field) at a pair of terminals (at a point):

$$Z_{in} = R_{in} + jX_{in} \qquad (1\text{-}42)$$

R_{in} and X_{in} are the input resistance and reactance of the antenna. The resistance is written as

$$R_{in} = R_r + R_L \qquad (1\text{-}43)$$

R_r and R_L are the radiation and the loss resistance of the antenna.

The antenna impedance can be derived by numerical methods and can be measured by well-tested techniques. It is noticed that the impedance is critical to the feed network design.

1.6 Antenna Arrays Classification

It is difficult for a single-element antenna to achieve narrow beams, low SLL, high directivity, high efficiency, null constraints, and so on due to its limited performance. A group of antenna elements, called an 'antenna array', is expected to attain the above desired performance. The types of elements, the geometry and the excitation characterize an antenna array. Knowledge of the mutual impedance between any pair of elements of the array, and the impedance of each one of them help to realize the feed network of the antenna. The network must be fairly simple to build and its size should occupy minimum space.

For simplicity in design and fabrication, the elements of an array are chosen in such a way so as to be identical and parallel. They must also be of lightweight at a relatively low cost. To meet certain performance requirements of the array, the physical space, the excitation distribution as well as the orientation and the type of elements must be derived. It is noticed that nowadays one can find a plethora of arrays. These range from complex electronically steered arrays to simpler static linear ones. A grouping of the usual antenna arrays depending on the form of their radiation pattern, their geometry, their type of elements and their excitation/fabrication is given below:

Radiation pattern: *broadside, end-fire, intermediate, omnidirectional, scanned, shaped beam, multiple beam, low SLL, adaptive, null constrained.*

Geometry: *linear, planar, circular, flat, three dimensional, conformal.*

Elements: *dipoles, monopoles, loops, slots, microstrip patches, horns, spiral, helices, log periodic, monolithic, active, electrically small.*

Excitation/fabrication: *uniform, binomial, Chebyshev, Fourier, Franklin, Hansen-Woodyard, even/odd mode, series fed, taper, phased, parasitic, magnetically scanned, signal processing, superconducting, monolithic, digital beam forming.*

1.7 Array Factor Classification

An arbitrary array with N elements produces a radiation pattern that, in general, can be characterized by the superimposition of the electric fields of its elements. From (1-3) and

(1-6), the electric field of the array is given by

$$E(r) = -j\omega\mu \frac{e^{-j\beta r}}{4\pi r} \sum_{n=1}^{N} I_n f_n(\theta, \phi)e^{j\beta r_n \cos \xi_n} \qquad (1\text{-}44)$$

$f_n(\theta, \phi)$ is the pattern function (vector) of the nth element.

As stated, the expression (1-44) contains the element patterns in the presence of the whole array. Obviously, the patterns for each element of the array are different. Even if the elements are identical and parallel, there must be some difference, which increases towards the array edge. However, as it was mentioned in 1.2, all the patterns of the elements of such an array are supposed to be approximately the same. In other cases, the radiation pattern is represented by the 'vector array pattern' and the corresponding array factor will be

$$AF(\theta, \phi) = \sum_{n=1}^{N} I_n f_n(\theta, \phi)e^{j\beta r_n \cos \xi_n} \qquad (1\text{-}45)$$

The above expression derives the vector array factor, which cannot be directly related to the array factor of isotropic sources. It is noticed that the synthesis of a general array represented in (1-45) is different than the corresponding one with a scalar factor.

References

[1] S.A. Schelkunoff and H.T. Friis, *Antenna Theory and Practice*, John Wiley & Sons, New York, 1952.

[2] R.W.P. King, *The Theory of Linear Antennas*, Harvard University Press, Cambridge, MA, 1956.

[3] J.D. Kraus, *Antennas*, McGraw-Hill Book Company, New York, 1988.

[4] R.C. Hansen (Ed.), *Microwave Scanning Antennas*, Academic Press, Vol. I, 1964, Vols. II & III, 1966, Peninsula Publishing, New York, 1985.

[5] R.S. Elliot, *Antenna Theory and Design*, Prentice Hall, Englewood Cliffs, NJ, 1981.

[6] C.A. Balanis, *Antenna Theory, Analysis and Design*, 3rd ed., John Wiley & Sons, New York, 2005.

[7] T. Milligan, *Modern Antenna Design*, McGraw-Hill Book Company, New York, 1985.

[8] N. Amitay, V. Galindo and C.P. Wu, *Theory and Analysis of Phased Arrays*, Wiley-Interscience, New York, 1972.

[9] M.T. Ma, *Theory and Applications of Antenna Arrays*, Wiley-Interscience, New York, 1974.

[10] A.W. Rudge, K. Milne, A.D. Olver and P. Knight (Eds.), *The Handbook of Antenna Design*, IEE/Peter Peregrinus Limited, London, 1983.

[11] Y.T. Lo and S.W. Lee, *Antenna Handbook*, Van Nostrad Reinhold, New York, 1988.

[12] R.C. Johnson and H. Jasik, *Antenna Engineering Handbook*, McGraw-Hill Book Company, New York, 1993.

[13] R.C. Mailloux, *Phased Array Antenna Handbook*, Artech House, Norwood, MA, 1994.

[14] J.R. James and P.S. Hall (Eds.), *Handbook of Microstrip Antennas*, Vol. I & II, IEE/Peter Peregrinus Limited, London, 1989.

[15] N. Fourikis, *Phased Array-Based Systems and Applications*, Wiley-Interscience, New York, 1997.

[16] R.C. Hansen, *Phased Array Antennas*, Wiley-Interscience, New York, 1998.

[17] R.T. Compton Jr., *Adaptive Antennas*, Prentice Hall, Englewood Cliffs, NJ, 1988.

[18] T.S. Rappaport (Ed.), *Smart Antennas*, IEEE Press, Piscataway, NJ, 1998.

[19] G.V. Tsoulos (Ed.), *Adaptive Antennas for Wireless Communications*, IEEE Press, Piscataway, NJ, 2001.

[20] Y. Rahmat-Samii and E. Michielssen, *Electromagnetic Optimization by Genetic Algorithms*, Wiley-Interscience, New York, 1999.

[21] B.D. Popovic, M.B. Dragovic and A.R. Djordjevic, *Analysis and Synthesis of Wire Antennas*, Research Studies Press, John Wiley & Sons, New York, 1982.

[22] W.L. Stutzman and G.A. Thiele, *Antenna Theory and Design*, John Wiley & Sons, New York, 1998.

[23] C.A. Balanis, *Advanced Engineering Electromagnetics*, John Wiley & Sons, New York, 1989.

[24] B.D. Popovic and B.M. Kolundzija, *Analysis of Metallic Antennas and Scatterers*, IEE Electromagnetic Waves Series IEE, Vol. **38**, London, 1994.

[25] A. Kumar, *Antenna Design with Fiber Optics*, Artech House, Norwood, MA, 1996.

[26] L.C. Godara, *Handbook of Antennas in Wireless Communications*, CRC Press, New York, 2002.

[27] J.D. Irwin, *Basic Engineering Circuit Analysis*, 2nd ed., Chap. 15, Macmillan, New York, 1987.

[28] V.H. Rumsey, "The Reaction Concept in Electromagnetic Theory," *Phys. Rev.*, Series 2, Vol. **94**, No. 6, pp. 1483–1491, June 15, 1954.

[29] H.G. Booker, V.H. Rumsey, G.A. Deschamps, M.I. Kales and J.I. Bonhert, "Techniques for Handling Elliptically Polarized Waves with Special Reference to Antennas," *Proc. IRE*, Vol. **39**, pp. 533–552, May 1951.

[30] W.H. Kummer and E.S. Gillespie, "Antenna Measurements-1978," *Proc. IEEE*, Vol. **66**, No. 4, pp. 483–507, April 1978.

[31] W.L. Stutzman, *Polarization in Electromagnetic Systems*, Artech House, Boston, MA, 1993.

[32] D.K. Cheng and F.I. Tseng, "Elliptical Arrays," *Proc. IEEE*, Vol. **114**, pp. 589–594, May 1967.

[33] K.R. Carver and J.W. Mink, "Microstrip Antenna Technology," *IEEE Trans. Antennas Propag.*, Vol. **AP-29**, No. 1, pp. 2–24, January 1981.

[34] E.N. Gilbert and S.P. Morgan, "Optimum Design of Directive Antenna Arrays Subject to Random Variations," *Bell Syst. Tech. J.*, Vol. **34**, pp. 637–663, May 1955.

2

Arrays: Linear, Planar, 3D and Conformal

2.1 Introduction

In this chapter, an analysis is given of the most common arrays. We start with linear arrays with uniform excitations and we next present arrays with Chebyshev, Taylor and Bayliss distributions. In planar arrays, the rectangular, the circular and the ring arrays are given. Next, 3-D arrays are analysed and some examples of cylindrical ones are presented. Finally, the characteristics of conformal arrays and several applications are discussed.

2.2 Linear Arrays

One of the methods to obtain directive antennas is to use several individual elements that add their contributions to preferred directions and cancel them in others. The above arrangement is known as an 'array'. The most practical array that consists of a number of identical and parallel elements subject to a straight line is the linear array.

Consider a typical linear array placed along the z-axis as shown in Fig. 2.1.

The array factor $AF(\theta, \phi)$ depends only on the angle θ and is written as

$$AF(\theta) = \sum_{n=0}^{N} I_n e^{j\beta d_n \cos\theta} \tag{2-1}$$

If the elements are positioned in the same inter-element distance d, then Equation (2-1) yields

$$AF(\theta) = \sum_{n=0}^{N} I_n e^{j\beta n d \cos\theta} = \sum_{n=0}^{N} I_n z^n \tag{2-2}$$

where

$$z = e^{j\beta d \cos\theta} \tag{2-3}$$

For $0 \le \theta \le \pi$, $AF(\theta)$ is a polynomial of z, which moves on a unit circle with a phase bounded between $-\beta d$ and $+\beta d$. The bounded region is called a 'visible region'. The unit

---

Orthogonal Methods for Array Synthesis: Theory and the ORAMA Computer Tool   John N. Sahalos
© 2006 John Wiley & Sons, Ltd

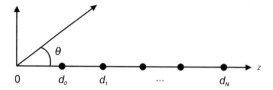

Figure 2.1   Geometry of a typical linear array

circle approach, proposed by Schelkunoff [1] gives a visual display of how the element contributions combine.

A linear array that cancels noise from N different directions has a pattern of the following form:

$$AF(\theta) = C \prod_{n=1}^{N}(z - z_n) = C \sum_{k=0}^{N} I_k z^k \qquad (2\text{-}4)$$

C is the normalization factor such that $|AF(\theta)|_{\max} = 1$. $z_n = e^{j\beta d \cos \theta_n}$, where θ_n is the direction of the nth interference and I_k is the required excitation of the kth element. To illustrate the pattern, let us consider an example. Suppose that a linear array with $N = 3$ elements and $d = \lambda/2$ should have nulls at $\theta_1 = 30°$ and $\theta_2 = 130°$.

From Equation (2-4), it is found that $I_0 = 0.6554 + j0.2386$, $I_1 = 1.0 + j0.0$ and $I_2 = 0.6554 - j0.2386$. The polar plot of the pattern is given in Fig. 2.2.

The type and orientation of the elements of the array are selected for receiving the maximum of the desired signal.

By varying z_n, it is possible to steer the nulls. This will alter the element excitations. In the example of Fig. 2.2, if the desired null moves from $130°$ to $120°$, the excitations will

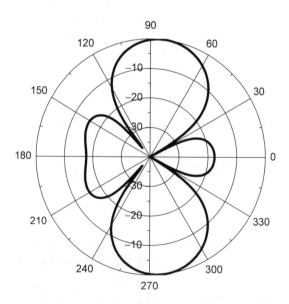

Figure 2.2   Polar plot of a pattern with nulls at $\theta_1 = 30°$ and $\theta_2 = 130°$

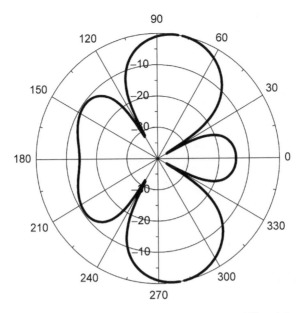

Figure 2.3   Polar plot of a pattern with nulls at $\theta_1 = 30°$ and $\theta_2 = 120°$

change to $I_0 = 0.7716 + j0.50$, $I_1 = 1.0 + j0.0$ and $I_2 = 0.7716 - j0.50$. The pattern is presented in Fig. 2.3.

A linear array with all roots equal is known as a 'binomial array'. $AF(\theta)$ for a binomial array is of the form

$$AF(\theta) = C(z - z_1)^N = C \sum_{k=0}^{N} \binom{N}{k} (-z_1)^k z^{N-k} \tag{2-5}$$

where

$$\binom{N}{k} = \frac{N!}{(N-k)!k!} \tag{2-6}$$

The binomial array has low minor lobes. However, a wide and difficult to realize variation among the amplitudes of the elements is present. This variation increases as the number of elements increases.

2.3 Uniform Linear Arrays

Linear arrays with equally spaced elements, identical magnitude and progressive phase α, are referred to as 'uniform arrays'. For N elements, we have $I_k = (e^{j\alpha})^k$. Thus, Equation (2-4) becomes

$$AF(\theta) = C \sum_{n=0}^{N-1} (ze^{j\alpha})^n = C \frac{(ze^{j\alpha})^N - 1}{(ze^{j\alpha}) - 1} \tag{2-7}$$

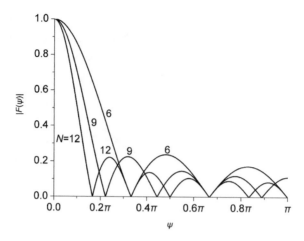

Figure 2.4   Array factors $|F(\psi)|$ versus ψ of uniform linear arrays with $N = 6, 9$ and 12 elements

According to [2], Equation (2-7) may be transformed into

$$AF(\theta) = e^{j(N-1)\psi/2} \frac{\sin(N\psi/2)}{N \sin(\psi/2)} \tag{2-8}$$

where

$$\psi = \beta d \cos \theta + \alpha \tag{2-9}$$

d is the inter-element distance of the array.

$|AF(\theta)|$ as a function of ψ, known also as $|F(\psi)|$, is presented for a few values of N in Fig. 2.4.

The main characteristics of $F(\psi)$ are the following:

1. Maximum values occur at $\psi = \pm 2k\pi$, where $k = 0, 1, 2, \ldots$
2. Nulls of the array are at $\psi = \pm 2k\pi/N$, where $k = 1, 2, 3, \ldots$ and $k \neq N, 2N, 3N, \ldots$
3. The 3 dB points of the array factor are at ψ, which gives

$$F(\psi) = \pm \frac{\sqrt{2}}{2} \Rightarrow |F(\psi)|^2 = \frac{1}{2} \Rightarrow 10 \log |F(\psi)|^2 = 3 \text{ dB}$$

Therefore, $N\psi/2 = \pm 1.391$ rad.

4. Secondary maxima (minor lobes) occur when

$$\psi = (2k+1)\pi/N,$$

where $k = 1, 2, 3, \ldots$

It is noticed that the first minor lobe is at $\psi = \pm 3\pi/N$, which gives

$$|F(\psi)| = \frac{1}{N \sin \dfrac{3\pi}{2N}} = SLL \tag{2-10}$$

When a beam maximum appears at $\theta = \pi/2$, the array is a broadside one. An end-fire array is an array with a beam maximum at $\theta = 0$ and/or π. The array beam can be steered in any direction θ_0 if the phase shift is $\alpha = -\beta d \cos \theta_0$. For an end-fire array, this is $\alpha = \mp \beta d$. In addition to the ordinary end-fire, there is the Hansen-Woodyard (HW) end-fire array, which is more directive [2]. In HW, the progressive phase shift is

$$\alpha = \mp \left(\beta d + \frac{2.94}{N} \right) \cong \mp \left(\beta d + \frac{\pi}{N} \right) \tag{2-11}$$

Moreover,

$$|\psi| = \pi \begin{cases} \text{for } \theta = \pi \text{ if maximum occurs at } \theta = 0 \\ \text{for } \theta = 0 \text{ if maximum occurs at } \theta = \pi \end{cases} \tag{2-12}$$

Condition (2-12) gives

$$d = \frac{\lambda}{4} \left(1 - \frac{2.94}{\pi N} \right) \cong \frac{\lambda}{4} \left(1 - \frac{1}{N} \right) \tag{2-13}$$

HW is useful for very long arrays with small inter-element distance. Useful formulas for the prescribed linear arrays are given in Table 2.1.

2.4 Chebyshev Linear Arrays

2.4.1 Chebyshev Polynomials

Chebyshev arrays are uniformly spaced linear arrays with non-uniform excitation. They make use of the Chebyshev polynomials [3].

The Chebyshev polynomial $T_m(x)$ of an independent variable x is an orthogonal polynomial of the mth order. It contains equal ripples in the region $-1 \leq x \leq 1$ and the amplitude varies between $+1$ and -1. The polynomial outside the above region rises exponentially. That is,

$$T_m(x) = \begin{cases} \cos(m \cos^{-1} x) & |x| \leq 1 \\ \left(\dfrac{x}{|x|} \right)^m \cosh(m \cosh^{-1} |x|) & |x| > 1 \end{cases} \tag{2-14}$$

and

$$\left. \begin{array}{l} T_0(x) = 1 \\ T_1(x) = x \end{array} \right\} \tag{2-15}$$

Equation (2-15) and the recursion relation

$$T_m(x) = 2x T_{m-1}(x) - T_{m-2}(x) \tag{2-16}$$

creates the Chebyshev polynomials of any order.

2.4.2 Dolph-Chebyshev Arrays

Dolph [4] recognized that Chebyshev polynomials could be used to have arrays with maximum directivity for a given side-lobe level. The equal ripples of the polynomials represent the side lobes, while the main beam comes from the exponential increase beyond $|x| = 1$.

The linear array is fed symmetrically to the central line (see Fig. 2.5).

The array factor can be expressed in terms of $\cos(\psi/2)$, where $\psi = \beta d \cos \theta + \alpha$.

The independent variable of the Chebyshev polynomial is

$$x = x_0 \cos(\psi/2) \tag{2-17}$$

Table 2.1 Formulas for uniform linear arrays

	Broadside	End fire	Hansen-Woodyard end fire	Intermediate with maximum at $\theta = \theta_0$
Directivity	$\sim 2N^{d}/_{\lambda}$	$\sim 4N^{d}/_{\lambda}$	$\sim 1.789(4N^{d}/_{\lambda})$	Depending on θ_0
HPBW	$\pi - 2\cos^{-1}\left(\dfrac{2.782}{N\beta d}\right)$	$2\cos^{-1}\left(1 - \dfrac{2.782}{N\beta d}\right)$	$2\cos^{-1}\left(1 - \dfrac{0.2796\pi}{N\beta d}\right)$	$\left\lvert \cos^{-1}\left(\cos\theta_0 + \dfrac{2.782}{N\beta d}\right) - \cos^{-1}\left(\cos\theta_0 - \dfrac{2.782}{N\beta d}\right)\right\rvert$
Beamwidth between nulls	$\pi - 2\cos^{-1}\left(\dfrac{2\pi}{N\beta d}\right)$	$2\cos^{-1}\left(1 - \dfrac{2\pi}{N\beta d}\right)$	$2\cos^{-1}\left(1 - \dfrac{\pi}{N\beta d}\right)$	$\left\lvert \cos^{-1}\left(\cos\theta_0 + \dfrac{2\pi}{N\beta d}\right) - \cos^{-1}\left(\cos\theta_0 - \dfrac{2\pi}{N\beta d}\right)\right\rvert$
Side lobe maximum positions	$\cos^{-1}\left(\pm\dfrac{(2k+1)\pi}{N\beta d}\right)$	$\cos^{-1}\left(1 - \dfrac{(2k+1)\pi}{N\beta d}\right)$ $k = 1,2,3,\ldots$	$\cos^{-1}\left(1 - \dfrac{2k\pi}{N\beta d}\right)$ $k \neq N, 2N, 3N, \ldots$	$\cos^{-1}\left(\cos\theta_0 \pm \dfrac{(2k+1)\pi}{N\beta d}\right)$
Null angular positions	$\cos^{-1}\left(\pm\dfrac{2k\pi}{N\beta d}\right)$	$\cos^{-1}\left(1 - \dfrac{2k\pi}{N\beta d}\right)$ $k = 1,2,3,\ldots$	$\cos^{-1}\left[1 + (1-2k)\dfrac{\pi}{N\beta d}\right]$ $k \neq N, 2N, 3N, \ldots$	$\cos^{-1}\left(\cos\theta_0 \pm \dfrac{2k\pi}{N\beta d}\right)$

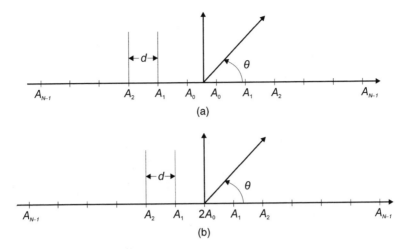

Figure 2.5   Linear array with symmetrical feed (a) N even and (b) N odd number of elements

At $x = x_0$, the polynomial has maximum value R. That is,

$$T_m(x_0) = R \quad \text{or} \quad x_0 = \cosh\left(\frac{1}{m}\cosh^{-1}R\right) \tag{2-18}$$

The zeros of $T_m(x)$ are at

$$x_k = \pm\cos\frac{(2k-1)\pi}{2m} \qquad k = 1, 2, \ldots, m \tag{2-19}$$

which correspond to

$$\psi_k = \pm 2\cos\left(\frac{x_k}{x_0}\right) \tag{2-20}$$

The excitations I_k are found from Equation (2-4) by using $z_k = e^{j\psi_k}$. The order of the Chebyshev polynomial is one less than the total number of elements of the array.

An example of a 12-element array with SLL $= -25$ dB is given. The polynomial is the $T_{11}(x)$. For $R = 25$ dB, it is $R = 10^{25/20} = 17.7828$ and from Equation (2-18), $x_0 = 1.0531$. A broadside array has the excitation coefficients presented in Table 2.2.

An intermediate array with maximum at $\theta = \theta_0$ has the same amplitude as before, with a phase shift $\alpha = -\beta d \cos\theta_0$. Figure 2.6 shows the patterns for $d/\lambda = 0.5$ of a broadside and an intermediate array with $\theta_0 = 60°$.

Table 2.2   Excitations of a broadside Dolph-Chebyshev array with $N = 12, d = 0.5\lambda$ and SLL $= -25$ dB

Element number	1,12	2,11	3,10	4,9	5,8	6,7
Excitation coefficient	0.4225	0.4572	0.6372	0.8031	0.9307	1.0000

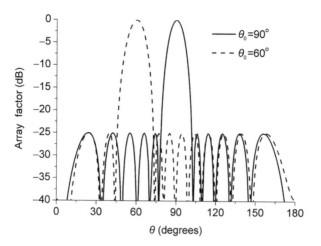

Figure 2.6 Patterns of Dolph-Chebyshev arrays with $N = 12$, SLL $= -25$ dB $d = 0.5\lambda$ and maximum at $\theta_0 = 90°$ and $\theta_0 = 60°$

2.4.3 Riblet Arrays

Dolph-Chebyshev arrays are suitable for $d \geq \lambda/2$ and fail to give the optimum design for $d < \lambda/2$. Riblet [5] devised a method to overcome the problem. He used $(2m + 1)$ elements and an independent variable of the form

$$x = a \cos \psi + b \tag{2-21}$$

The polynomial is $T_m(x)$ with a visible region $-1 \leq x \leq x_0$. It is

$$\left. \begin{array}{l} x_0 = a + b \\ -1 = a \cos \beta d + b \end{array} \right\} \tag{2-22}$$

By solving Equation (2-22), we have

$$a = \frac{1 + x_0}{1 - \cos \beta d} \text{ and } b = -\frac{1 + x_0 \cos \beta d}{1 - \cos \beta d} \tag{2-23}$$

Zeros of $T_m(x)$ are given by Equation (2-19) and the corresponding ψ_k are

$$\psi_k = \pm \cos^{-1} \left(\frac{x_k - b}{a} \right) \tag{2-24}$$

An 11 element array with $d/\lambda = 0.4$ and SLL $= -20$ dB is presented. The excitation is given in Table 2.3 and the pattern in Fig. 2.7.

Table 2.3 Excitations of a broadside Riblet array with $N = 11, d/\lambda = 0.4$ and SLL $= -20$ dB

Element number	1,11	2,10	3,9	4,8	5,7	6
Excitation coefficient	0.4776	−0.0910	0.7574	−0.1771	1.0000	−0.2149

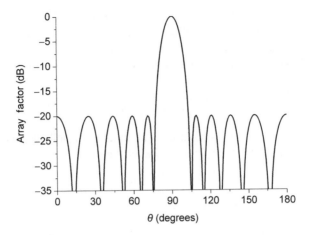

Figure 2.7 Pattern of a Riblet array with $N = 11, d/\lambda = 0.4$ and $SLL = -20$ dB

2.4.4 Other Chebyshev Arrays

The Riblet technique can be extended to the design of linear Chebyshev arrays in a more general form.

We start with the independent variable x, which is written as

$$x = a\cos(\beta d\cos\theta + \alpha) + b \tag{2-25}$$

To find a, b and α, three characteristic angles θ_1, θ_2 and θ_3, which correspond to three values x_1, x_2 and x_3 of x, must be defined:

$$\left.\begin{aligned}
x_1 &= a\cos(\beta d\cos\theta_1 + \alpha) + b \\
x_2 &= a\cos(\beta d\cos\theta_2 + \alpha) + b \\
x_3 &= a\cos(\beta d\cos\theta_3 + \alpha) + b
\end{aligned}\right\} \tag{2-26}$$

By solving Equation (2-26), one can find

$$\tan\alpha = \frac{\sin y_{21} - \lambda\sin y_{31}}{\lambda\cos y_{31} - \cos y_{21}} \tag{2-27}$$

where

$$\left.\begin{aligned}
y_{21} &= \beta\frac{d}{2}(\cos\theta_2 + \cos\theta_1) \\
y_{31} &= \beta\frac{d}{2}(\cos\theta_3 + \cos\theta_1) \\
\lambda &= \frac{(x_2 - x_1)\sin\left[\dfrac{\beta d}{2}(\cos\theta_3 - \cos\theta_1)\right]}{(x_3 - x_1)\sin\left[\dfrac{\beta d}{2}(\cos\theta_2 - \cos\theta_1)\right]}
\end{aligned}\right\} \tag{2-28}$$

$$a = \frac{x_2 - x_1}{\cos(\beta d\cos\theta_2 + \alpha) - \cos(\beta d\cos\theta_1 + \alpha)} \tag{2-29}$$

$$b = x_1 - a\cos(\beta d\cos\theta_1 + \alpha) \tag{2-30}$$

Table 2.4 and Figs. 2.8–2.11 present four common end-fire cases coming from Equations (2-26–2-30). It is noticed that, with the above expressions, one can produce various broadside, end fire and intermediate Chebyshev arrays. A general approach to Chebyshev and Legendre arrays, where Legendre polynomials are used, can be found in [6, 7].

Array factors of the form

$$AF(\theta) = T_m(x)T_1^n(x) \tag{2-31}$$

can give either equal or unequal side lobes. The number of nulls depends on m and n. If we compare Equation (2-31) with an array factor then in

$$AF_1(\theta) = T_{m+n}(x) \tag{2-32}$$

we see that $AF_1(\theta)$ has $(m+n)$ roots while $AF(\theta)$ has either $(m+1)$ roots for $m = 2k$ or m roots for $m = 2k+1$.

Figure 2.12 presents for comparison the factors $T_5(x)$ and $T_3(x)T_1^2(x)$ of the case 3 end-fire array for $N = 11, d/\lambda = 0.35$ and SLL $= -20$ dB. Also, the same array with factors $L_5(x)$ and $L_3(x)L_1^2(x)$ is given in Fig. 2.13.

From the Figs. 2.8–2.12, it is shown that the minimum HPBW, as it is expected, comes from the $T_5(x)$ pattern. All the other patterns, due to the unequal side lobes, have larger values of HPBW.

2.5 Linear Arrays from Sampling or Root Matching of Line Sources

Continuous distributions [2, 8, 9] create excellent patterns with low side lobes. Discrete arrays, coming from sampling of continuous distributions, under certain criteria, can give similar patterns. Where there is large element spacing, the sampling criterion is not valid [10] and the patterns of the array and the line source do not match well. A method of root matching, where the nulls of the pattern of the continuous distribution appear in the pattern of the discrete array, is considered. If the pattern does not yield the desired accuracy, a perturbation technique [11] can be applied. In this case, the distribution of the discrete-element array varies to improve accuracy.

2.5.1 Simple Linear Distributions

A discrete-element array of fixed length is transformed into a line source as the number of elements approaches infinity (Fig. 2.14). The array factor reduces to an integral and is called a 'space factor' (*SF*):

$$SF(\theta) = \int_{-L/2}^{L/2} I(z')e^{j(\beta\cos\theta-\alpha)z'} \, dz' = \int_{-L/2}^{L/2} I(z')e^{j\xi z'} \, dz' \tag{2-33}$$

$I(z')$ and α are the amplitude distribution and phase progress along the source.

Equation (2-33) is the finite 1-D Fourier transform that relates to the far field with the excitation.

Changing the bounds of integration, Equation (2-33) becomes

$$SF(\theta) = \int_{-\infty}^{\infty} I(z')e^{j\xi z'} \, dz' \tag{2-34}$$

Table 2.4 Coefficients of Chebyshev end-fire arrays

End-fire case 1	End-fire case 2	End-fire case 3 (optimum)	End-fire case 4
$d_{\max} = \lambda/4$	$d_{\max} = \lambda/4$	$d_{\max} = \dfrac{\lambda}{2} - \dfrac{\lambda}{2\pi}\sin^{-1}\left(\sqrt{\dfrac{x_1 + x_3}{x_1 + 1}}\right)$	$d_{\max} = \dfrac{\lambda}{2\pi}\left[\alpha - \sin^{-1}\left(\sqrt{\dfrac{x_1 + x_2}{x_1 + 1}}\right)\right]$
$x_1 \to \theta_1 = 0$	$x_1 \to \theta_1 = 0$	$x_1 \to \theta_1 = 0$	$x_1 \to \theta_1 = 0$
$x_2 \to \theta_2 = \pi$	$x_2 \to \theta_2 = \pi$	$x_2 \to \theta_2 = \pi$	$x_2 \to \theta_2 = \pi$
$x_3\#$	$x_3\#$	$x_3 = -(a+b)$	$x_3 \to \theta_3 = \theta_{HP}$
$T_m(x_1) = R$	$T_m(x_1) = R$	$T_m(x_1) = R$	$T_m(x_1) = R,\; T_m(x_3) = R\sqrt{2}/2$
$a = \dfrac{x_1 + x_2}{1 - \cos 2\beta d}$	$a = \dfrac{x_1 - x_2}{\cos 2\beta d - 1}$	$a = \dfrac{x_2 - x_1}{2\sin\beta d \cdot \sin\alpha}$	$a = \dfrac{x_1 + x_2}{2\sin\beta d \cdot \sin\alpha}$
$b = x_1 - a$	$b = x_2 - a$	$b = -x_3 - a$	$b = x_1 - \cos(\beta d + \alpha)$
$\alpha = -\beta d$	$\alpha = \beta d$	$\alpha = 2\tan^{-1}\left[\sin(\beta d)\dfrac{x_1 + x_2 + 2x_3 - 2\sqrt{(x_1 + x_3)(x_2 + x_3)}}{(x_1 - x_2)[1 + \cos(\beta d)]}\right]$	$\alpha = \cot^{-1}\left[\dfrac{(2\lambda + 1)\sin\beta d - \sin(\beta d\cos\theta_3)}{\cos\beta d - \cos(\beta d\cos\theta_3)}\right]$
			$\lambda = \dfrac{x_3 - x_1}{x_1 + x_2}$

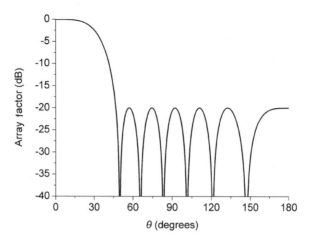

Figure 2.8   Pattern of an end-fire (case 1) array with $N = 13$, $d/\lambda = 0.25$, SLL $= -20$ dB

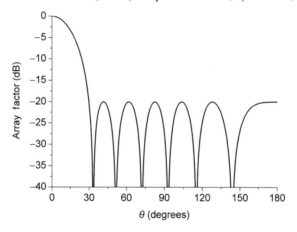

Figure 2.9   Pattern of an end-fire (case 2) array with $N = 13$, $d/\lambda = 0.20$, SLL $= -20$ dB

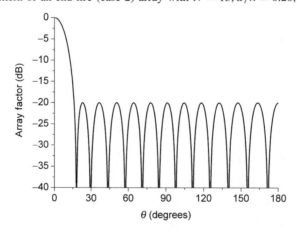

Figure 2.10   Pattern of an end-fire (case 3) array with $N = 13$, $d/\lambda = 0.25$, SLL $= -20$ dB

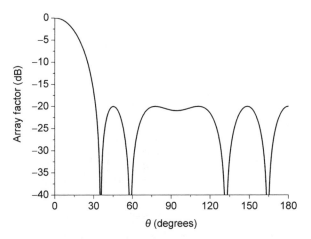

Figure 2.11 Pattern of an end-fire (case 4) array with $N = 13$, $d/\lambda = 0.30$, SLL $= -20$ dB and HPBW $= 30°$

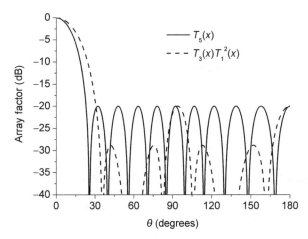

Figure 2.12 Patterns of $T_5(x)$ and $T_3(x)T_1^2(x)$ (case 3) for $N = 11$ $d/\lambda = 0.35$ and SLL $= -20$ dB

$I(z')$ is zero outside of $-L/2 \leq z' \leq L/2$. By using the Fourier transform, we have

$$I(z') = \frac{1}{2\pi} \int_{-\infty}^{\infty} SF(\theta)e^{-jz'\xi} \, d\xi \qquad (2\text{-}35)$$

If L is large enough Equation (2-34) gives the desired pattern within a certain margin of error. In the discrete-element array, $I(z')$ is sampled at appropriate intervals. Useful distributions with their characteristics are presented in Table 2.5.

2.5.2 Taylor Distribution (Chebyshev Error)
Chebyshev arrays provide an optimum relation between the SLL and the HPBW. Another distribution characterized by low SLL of the first N side lobes next to the main lobe is the Taylor distribution. The rest of the side lobes gradually fall off in value. The SF of

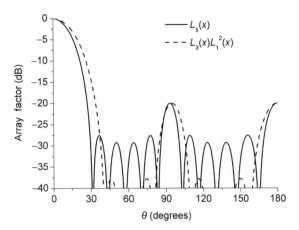

Figure 2.13 Patterns of $L_5(x)$ and $L_3(x)L_1^2(x)$ (case 3) for $N = 11 d/\lambda = 0.35$ and SLL = -20 dB

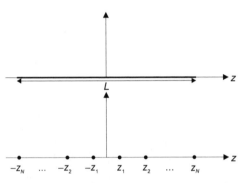

Figure 2.14 Line source and its equivalent discrete-element array

the Taylor distribution comes from the Dolph-Chebyshev distribution if the elements of the array become infinite [8,12]:

$$SF(\theta) = \frac{\cosh\left[\sqrt{(\pi A)^2 - u^2}\right]}{\cosh(\pi A)} \tag{2-36}$$

where

$$\left.\begin{array}{c} u = \pi \dfrac{L}{\lambda}(\cos\theta - \cos\theta_0) \\[2mm] \cosh(\pi A) = R \end{array}\right\} \tag{2-37}$$

Since Equation (2-36) cannot be realized physically, Taylor [12] presented a SF whose roots are the zeros of $SF(\theta)$. Because the factor is the approximation of the ideal Chebyshev, it is known as the 'Chebyshev error factor'. The SF is

$$SF(u, A, \overline{n}) = \frac{\sin u}{u} \frac{\prod_{n=1}^{\overline{n}-1}\left[1 - \left(\dfrac{u}{u_n}\right)^2\right]}{\prod_{n=1}^{\overline{n}-1}\left[1 - \left(\dfrac{u}{n}\right)^2\right]} \tag{2-38}$$

Table 2.5 Radiation characteristics for line sources with uniform, triangular, cosine and cosine-squared distributions

Distribution	Distribution I_n	Space factor (SF) $u=\left(\frac{\pi L}{\lambda}\right)\sin\theta$	Half-power beamwidth (degrees) $L \gg \lambda$	First null beamwidth (degrees) $L \gg \lambda$	First side lobe max. (to main max.) (dB)	Directivity factor (L large)
Uniform	I_0	$I_0 L \dfrac{\sin u}{u}$	$\dfrac{50.6}{(L/\lambda)}$	$\dfrac{114.6}{(L/\lambda)}$	-13.2	$2\left(\dfrac{L}{\lambda}\right)$
Triangular	$I_1\left(1-\dfrac{2}{L}|z'|\right)$	$I_1\dfrac{L}{2}\left[\dfrac{\sin\left(\frac{u}{2}\right)}{\frac{u}{2}}\right]^2$	$\dfrac{73.4}{(L/\lambda)}$	$\dfrac{229.2}{(L/\lambda)}$	-26.4	$0.75\left[2\left(\dfrac{L}{\lambda}\right)\right]$
Cosine	$I_2\cos\left(\dfrac{\pi}{L}z'\right)$	$I_2 L \dfrac{\pi}{2}\dfrac{\cos(u)}{(\pi/2)^2-u^2}$	$\dfrac{68.8}{(L/\lambda)}$	$\dfrac{171.9}{(L/\lambda)}$	-23.2	$0.810\left[2\left(\dfrac{L}{\lambda}\right)\right]$
Cosine-squared	$I_3\cos^2\left(\dfrac{\pi}{L}z'\right)$	$I_3\dfrac{L}{2}\dfrac{\sin(u)}{u}\left[\dfrac{\pi^2}{\pi^2-u^2}\right]$	$\dfrac{83.2}{(L/\lambda)}$	$\dfrac{229.2}{(L/\lambda)}$	-31.5	$0.667\left[2\left(\dfrac{L}{\lambda}\right)\right]$

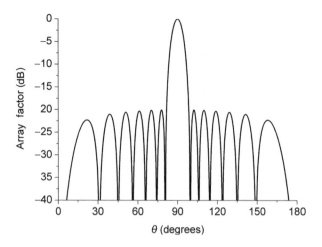

Figure 2.15   Taylor pattern for $L = 6.5\lambda$, $\bar{n} = 6$, SLL $= -20$ dB from an array with $N = 14$ and $d/\lambda = 0.5$

$(\bar{n} - 1)$ is a parameter, which defines the number of pairs of inner nulls. Moreover,

$$u_n = \bar{n}\frac{\sqrt{A^2 + \left(n - \dfrac{1}{2}\right)^2}}{\sqrt{A^2 + \left(\bar{n} - \dfrac{1}{2}\right)^2}} \qquad n = 1, 2, \ldots, \bar{n} - 1 \qquad (2\text{-}39)$$

The Taylor distribution of the line source can be found by the Fourier transform

$$I(z') = SF(0, A, \bar{n}) + 2\sum_{m=1}^{\bar{n}-1} SF(m, A, \bar{n}) \cos\left(m\pi\frac{z'}{L}\right) \qquad (2\text{-}40)$$

Sampling $I(z')$, we create a discrete linear array. Problems which arise and cause inaccuracies even for large arrays have been addressed earlier [13].

Figure 2.15 presents the Taylor pattern for SLL $= -20$ dB, $L = 6.5\lambda$ and $(\bar{n} = 6)$ taken from an array with $N = 14$ elements and $d/\lambda = 0.5$. The relative amplitudes of the elements of the array are given in Fig. 2.16.

2.5.3 Taylor One-parameter Distribution

In low noise systems, it is desirable to have the first side lobes at a certain level while the others decay as the angle increases. Taylor [2] developed a procedure to create such a pattern. The line source distribution is referred to as the 'Taylor one-parameter' and is of the following form:

$$I_n(z') = I_0\left[\pi B\sqrt{1 - \left(\frac{2z'}{L}\right)^2}\right] \qquad (2\text{-}41)$$

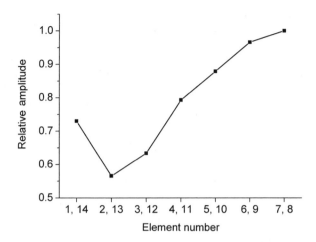

Figure 2.16 Relative amplitudes of the elements of the array corresponding to the pattern of Fig. 2.15

where
z' is the distance from the centre of the line source
L is the length of the line source
B is the parameter that determines the side lobe
I_0 is the modified Bessel function of the first kind and zero order [3].

The SF associated with Equation (2-41) can be obtained by the Fourier transform:

$$SF(\theta) = \begin{cases} \dfrac{\sin\left[\sqrt{(\pi B)^2 - u^2}\right]}{\sqrt{(\pi B)^2 - u^2}}, & u^2 < (\pi B)^2 \\ \dfrac{\sinh\left[\sqrt{u^2 - (\pi B)^2}\right]}{\sqrt{u^2 - (\pi B)^2}}, & u^2 > (\pi B)^2 \end{cases} \tag{2-42}$$

where

$$u = \beta \frac{L}{2}(\cos\theta - \cos\theta_0) \tag{2-43}$$

Parameter B is found from [14]:

$$R = 4.60333\frac{\sinh \pi B}{\pi B} \tag{2-44}$$

where R is the ratio of the beam peak to the side-lobe level.

2.5.4 Bayliss Distribution

A pattern null on boresight with the appropriate side-lobe level has been developed by Bayliss [15]. Monopulse tracking systems use an auxiliary pattern of the form of Bayliss coinciding with a beam peak of the main pattern.

Table 2.6 Coefficients of the parameters A, ξ_n and $u_{\max}$

x	A	ξ_1	ξ_2	ξ_3	ξ_4	$u_{\max}$
α_1	0.3038753	0.9858302	2.00337487	3.00636321	4.00518423	0.4797212
α_2	0.05042922	0.0333885	0.01141548	0.00683394	0.00501795	0.01456692
α_3	−0.00027989	0.00014064	0.0004159	0.00029281	0.00021735	−0.00018739
$\alpha_4 \cdot 10^5$	0.343	−0.19	−0.373	−0.161	−0.088	0.218
$\alpha_5 \cdot 10^7$	−0.2	0.1	0.1	0	0	−0.1

The Bayliss pattern is described in terms of two parameters A and $\bar{n}$:

$$SF(\theta) = u\cos(\pi u)\frac{\prod_{n=1}^{\bar{n}-1}\left[1-\left(\dfrac{u}{u_n}\right)^2\right]}{\prod_{n=1}^{\bar{n}-1}\left[1-\left(\dfrac{u}{n+\dfrac{1}{2}}\right)^2\right]} \tag{2-45}$$

$(\bar{n}-1)$ is the parameter, which defines the number of inner nulls. Moreover,

$$u_n = \begin{cases} \left(\bar{n}+\dfrac{1}{2}\right)\left(\dfrac{\xi_n^2}{A^2+\bar{n}^2}\right)^{1/2} & n = 1,2,3,4 \\ \left(\bar{n}+\dfrac{1}{2}\right)\left(\dfrac{A^2+n^2}{A^2+\bar{n}^2}\right)^{1/2} & n = 5,6,\ldots,\bar{n}-1 \end{cases} \tag{2-46}$$

A, ξ_n and the location $u_{\max}$, where SF is maximized, are found as a function of $S = |\text{side-lobe level(dB)}|$. It is (see Table 2.6)

$$x = \alpha_1 + S\{\alpha_2 + S[\alpha_3 + S(\alpha_4 + S\cdot\alpha_5)]\} \tag{2-47}$$

The aperture distribution by the sine Fourier series with $\bar{n}$ terms is

$$I(z') = \sum_{m=0}^{\bar{n}-1} B_m \sin\left[\left(m+\frac{1}{2}\right)\frac{\pi z'}{L}\right] \tag{2-48}$$

where

$$B_m = \frac{(-1)^m\left(m+\dfrac{1}{2}\right)^2\prod_{n=1}^{\bar{n}-1}\left[1-\left(\dfrac{m+\dfrac{1}{2}}{u_n}\right)^2\right]}{\prod_{n=1}^{\bar{n}-1}\left[1-\left(\dfrac{m+\dfrac{1}{2}}{n+\dfrac{1}{2}}\right)^2\right]} \tag{2-49}$$

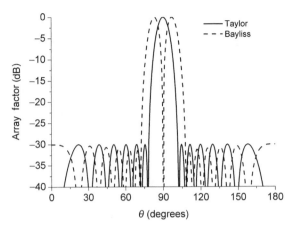

Figure 2.17 Bayliss and Taylor patterns ($\bar{n} = 7$) for SLL $= -30$ dB for a linear array with $N = 14$ and $d/\lambda = 0.5$

A Bayliss and a Taylor pattern ($\bar{n} = 7$) for SLL $= -30$ dB for an array with $N = 14$ and $d/\lambda = 0.5$ are presented in Fig. 2.17.

2.5.5 Modified Patterns by Iteration

By using a perturbation procedure, we can make Taylor and Bayliss patterns with individual different side lobes. According to Elliot [8], we express these patterns in more general forms. These are:

1. Taylor

$$SF(u) = C_0 \frac{\sin \pi u}{u} \frac{\prod_{n=-(\bar{n}_L-1)}^{\bar{n}_R-1} \left(1 - \dfrac{u}{u_n}\right)}{\prod_{n=-(\bar{n}_L-1)}^{\bar{n}_R-1} \left(1 - \dfrac{u}{n}\right)} \qquad (2\text{-}50)$$

where

$$\left. \begin{aligned} u_n &= \bar{n}_R \frac{\sqrt{A_R^2 + \left(n - \dfrac{1}{2}\right)^2}}{\sqrt{A_R^2 + \left(\bar{n}_R - \dfrac{1}{2}\right)^2}} \qquad n = 1, 2, \ldots, \bar{n}_R - 1 \\[2em] u_n &= -\bar{n}_L \frac{\sqrt{A_L^2 + \left(n + \dfrac{1}{2}\right)^2}}{\sqrt{A_L^2 + \left(\bar{n}_L - \dfrac{1}{2}\right)^2}} \qquad n = -1, -2, \ldots, -(\bar{n}_L - 1) \end{aligned} \right\} \qquad (2\text{-}51)$$

R and L identify the right and left side of the pattern. $\bar{n}_R$ and $\bar{n}_L$ denote the transition roots of the two sides and A_R and A_L are the corresponding SLL parameters.

2. Bayliss

$$SF(u) = C_0 u \cos \pi u \frac{\prod_{n=-(\bar{n}_L-1)}^{\bar{n}_R-1} \left(1 - \dfrac{u}{u_n}\right)}{\prod_{n=-\bar{n}_L}^{\bar{n}_R-1} \left(1 - \dfrac{u}{n + \dfrac{1}{2}}\right)} \tag{2-52}$$

We start with a pattern $SF_0(u)$ with the SLL on both sides being the average of the desired ones. All the roots of Equations (2-50) and (2-52) u_n^0 are known.

We assume that the roots of the desired pattern are

$$u_n = u_n^o + \delta u_n \tag{2-53}$$

δu_n is a small perturbation. If

$$C = C_0 + \delta C$$

$SF(u)$ becomes

$$\frac{SF(u)}{SF_0(u)} - 1 = \frac{\delta C}{C_0} + \sum_{n=-(\bar{n}_L-1)}^{\bar{n}_R-1} \frac{\dfrac{u}{(u_n^0)^2}}{1 - \dfrac{u}{u_n^0}} \delta u_n \tag{2-54}$$

The peak positions u_m^p in the Taylor distribution give

$$\frac{SF(u_m^p)}{SF_0(u_m^p)} - 1 = \frac{\delta C}{C_0} + \sum_{n=-(\bar{n}_L-1)}^{\bar{n}_R-1} \frac{\dfrac{u_m^p}{(u_n^0)^2}}{1 - \dfrac{u_m^p}{u_n^0}} \delta u_n \tag{2-55}$$

For the $\bar{n}_R + \bar{n}_L - 1$ lobes, we have an equal number of linear equations of the form Equation (2-55). It is noticed that $\delta u_0 = 0$ and that the system is solved for $\delta C/C_0$ and the $\bar{n}_R + \bar{n}_L - 2$ values δu_n. The new values of u_n are substituted in Equation (2-50) and the new pattern is checked. The process is repeated until the new pattern differs from the desired one by a minimum predefined amount.

The same procedure is applied for the Bayliss distribution. Equation (2-54) is modified to

$$\frac{SF(u_m^p)}{SF_0(u)} - 1 = \frac{\delta C}{C_0} - \frac{\delta u_0}{u_m^p} + \sum_{n=-(\bar{n}_L-1)}^{\bar{n}_R-1} \frac{\dfrac{u_m^p}{(u_n^0)^2}}{1 - \dfrac{u_m^p}{u_n^0}} \delta u_n \tag{2-56}$$

Equation (2-56) gives a system of $\bar{n}_R + \bar{n}_L$ unknowns, which is solved. The perturbation process is repeated in the same way.

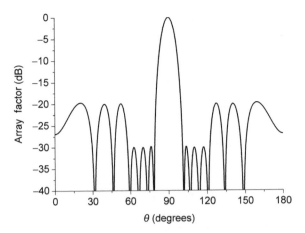

Figure 2.18   Pattern of the modified Taylor distribution for $\bar{n} = 6$ and SLL $= -20$ dB and the three innermost pairs of lobes -30 dB

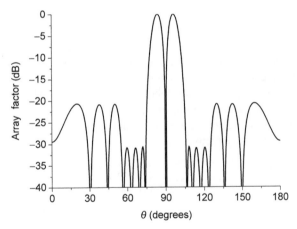

Figure 2.19   Pattern of the modified Bayliss distribution for $\bar{n} = 6$ and SLL $= -20$ dB and the three innermost pairs of lobes -30 dB

Figures 2.18 and 2.19 show the pattern of two modified distributions for $\bar{n} = 6$, SLL $= -20$ dB and three innermost pairs of lobes -30 dB.

In these cases, an iterative procedure can be applied for power pattern synthesis, where we have additional degrees of freedom. Orchard et al. [16] proposed a technique of dividing the pattern into two main regions (the shaped beam region and the side lobe region).

The array factor, in general, is

$$AF(\theta) = \prod_{n=1}^{N}(z - z_n) = \sum_{n=0}^{N} I_n z^n \qquad (2\text{-}57)$$

It is assumed that the zero locations are complex of the form

$$z_n = \exp(a_n + jb_n) \tag{2-58}$$

and z is written as

$$z = \exp(j\phi) \tag{2-59}$$

Orchard sets the Nth root $z_N = 1$ and expresses the power pattern in dB

$$G = \sum_{n=1}^{N-1} 10\log[1 - 2e^{a_n}\cos(\phi - b_n) + e^{2a_n}] + 10\log[2(1 + \cos\phi)] + C_1 \tag{2-60}$$

C_1 is a constant that allows G to have a given value at the main beam.

The unknown coefficients a_n, b_n and ϕ are found by using an iterative scheme. This scheme uses the derivatives of G and the difference between the existing and the desired power pattern. The procedure does not produce an optimum result. However, it offers flexibility and control of the ripple and the side-lobe level as well as of the entire radiation pattern.

2.6 Planar Arrays

Individual radiators positioned on a plane constitute a planar array. The usual planar array is rectangular. The elements are positioned along a rectangular grid. The grid can be made from the combination of two perpendicular linear arrays. Owing to the increase in the variables, the pattern can be controlled and scanned at any point in space. The more symmetrical patterns and the lower side lobes are the two main advantages of planar over linear arrays. In Fig. 2.20, a rectangular array is presented.

The element corresponding to the mth row and the nth column is the mnth element with excitation I_{mn}. The array factor is

$$AF(\theta, \phi) = \sum_{m=1}^{M}\sum_{n=1}^{N} I_{mn} z_x^m z_y^n \tag{2-61}$$

where

$$z_x = e^{j\beta d_x \sin\theta \cos\phi} \quad \text{and} \quad z_y = e^{j\beta d_y \sin\theta \sin\phi} \tag{2-62}$$

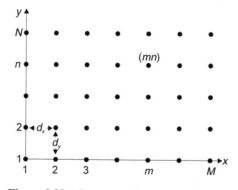

Figure 2.20   Geometry of a rectangular array

For uniform excitation and progressive phase α_x along x-axis and α_y along y-axis, the array factor becomes

$$AF(\theta, \phi) = I_0 \left(\sum_{m=1}^{M} e^{j(m-1)(\beta d_x \sin \theta \cos \phi + \alpha_x)} \right) \left(\sum_{n=1}^{N} e^{j(n-1)(\beta d_y \sin \theta \sin \phi + \alpha_y)} \right) \quad (2\text{-}63)$$

According to Equations(2-8) and (2-9), expression Equation (2-63) gives

$$|AF(\theta, \phi)| = \left| \frac{\sin(M\psi_x/2)}{M \sin(\psi_x/2)} \right| \left| \frac{\sin(N\psi_y/2)}{N \sin(\psi_y/2)} \right| \quad (2\text{-}64)$$

where

$$\left. \begin{array}{l} \psi_x = \beta d_x \sin \theta \cos \phi + \alpha_x \\ \psi_y = \beta d_y \sin \theta \sin \phi + \alpha_y \end{array} \right\} \quad (2\text{-}65)$$

It is obvious that, for d_x and/or $d_y \geq \lambda$, the in-phase addition of the radiated field is made in more than one directions and grating lobes are produced. The grating lobes are located at

$$\left. \begin{array}{l} \psi_x = \pm 2m\pi \\ \\ \psi_y = \pm 2n\pi \end{array} \quad m \text{ and } n = 1, 2, \ldots \right\} \quad (2\text{-}66)$$

If the main lobe direction is at (θ_0, ϕ_0), then

$$\left. \begin{array}{l} \alpha_x = -\beta d_x \sin \theta_0 \cos \phi_0 \\ \alpha_y = -\beta d_y \sin \theta_0 \sin \phi_0 \end{array} \right\} \quad (2\text{-}67)$$

By solving Equation (2-66) for the direction (θ_{mn}, ϕ_{mn}) where grating lobes occur, we have

$$\phi_{mn} = \tan^{-1} \left[\frac{\sin \theta_0 \sin \phi_0 \pm n\lambda/d_y}{\sin \theta_0 \cos \phi_0 \pm m\lambda/d_x} \right] \quad (2\text{-}68)$$

$$\theta_{mn} = \sin^{-1} \left[\frac{\sin \theta_0 \sin \phi_0 \pm n\lambda/d_y}{\sin \phi_{mn}} \right] = \sin^{-1} \left[\frac{\sin \theta_0 \cos \phi_0 \pm m\lambda/d_x}{\cos \phi_{mn}} \right] \quad (2\text{-}69)$$

An 8×6 element array with $\alpha_x = \alpha_y = 0$ will be given. Figures 2.21–2.23 show the corresponding patterns for $d_x = d_y = \lambda/2, d_x = \lambda/2, d_y = \lambda$ and $d_x = d_y = \lambda$. It is obvious that for large spacing, grating lobes appear.

2.6.1 Planar Chebyshev Arrays

In a planar array, instead of uniform excitation, one can have $AF(\theta, \phi)$ as a product of two different linear array factors of any permitted type. As an example, the three-dimensional pattern of an 8×8 rectangular array that combines two different Dolph-Chebyshev arrays with $SLL_x = -20$ dB for $d_x = 0.5\lambda$ and $SLL_y = -20$ dB for $d_y = 0.5\lambda$ is shown in Fig. 2.24. The corresponding planar patterns in four ϕ cuts are presented in Fig. 2.25. In the above case, it is evident that the side lobes do not have the same level on any plane. If it is desirable to have a planar array with side lobes of equal level in the three-dimensional pattern, we must follow a different approach [17, 18].

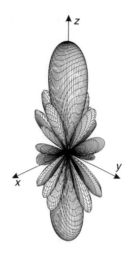

Figure 2.21 3-D pattern of an 8×6 uniform rectangular array with $\alpha_x = \alpha_y = 0$ and $d_x = d_y = \lambda/2$

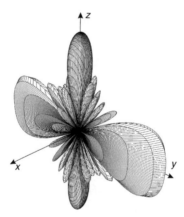

Figure 2.22 3-D pattern of an 8×6 uniform rectangular array with $\alpha_x = \alpha_y = 0$ and $d_x = \lambda/2$, $d_y = \lambda$

The desired array factor, instead of being a product of two Chebyshev factors, can be presented as

$$AF(\theta, \phi) = T_{N-1}(wxy) \qquad (2\text{-}70)$$

where

$$T_{N-1}(w) = R \qquad (2\text{-}71)$$

$$x = \cos\left(\beta\frac{d_x}{2}\sin\theta\cos\phi\right) \qquad (2\text{-}72)$$

$$y = \cos\left(\beta\frac{d_y}{2}\sin\theta\sin\phi\right) \qquad (2\text{-}73)$$

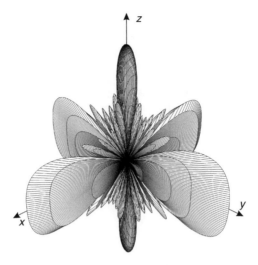

Figure 2.23 3-D pattern of an 8×6 uniform rectangular array with $\alpha_x = \alpha_y = 0$ and $d_x = d_y = \lambda$

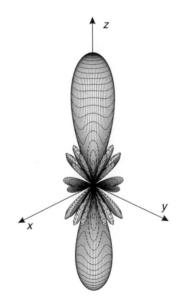

Figure 2.24 3-D pattern of an 8×8 Chebyshev rectangular array with $\alpha_x = \alpha_y = 0, d_x = d_y = \lambda/2$ and $SLL_x = SLL_y = -20$ dB

By equating the second members of Equations (2-61) and (2-70) for $M = N$ and for

$$I_{mn} = I_{-m,n} = I_{m,-n} = I_{-m,-n} \tag{2-74}$$

we can find the excitations of the elements of the array. Figure 2.26 presents the planar radiation patterns of the same 8×8 array, in four different cuts. It is observed that the side lobe is always of the same level.

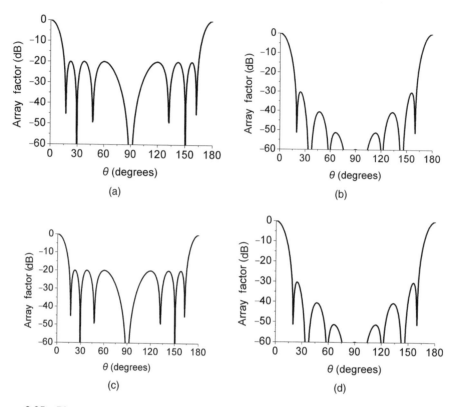

Figure 2.25 Planar patterns of the one given in Fig. 2.24 for ϕ equal to (a) 0°, (b) 30°, (c) 90° and (d) 120°

2.6.2 Circular Arrays

A circular array is a planar array with its elements positioned on a circular ring. A circular array with N isotropic elements (see Fig. 2.27) produces an array factor of the form

$$AF(\theta, \phi) = \sum_{n=1}^{N} I_n e^{j[\beta R \sin\theta \cos(\phi - \phi_n) + \alpha_n]} \tag{2-75}$$

I_n is the amplitude of the excitation of the nth element and α_n is the corresponding phase.

To have a peak at (θ_0, ϕ_0), it must be

$$\alpha_n = -\beta R \sin\theta_0 \cos(\phi_0 - \phi_n) \tag{2-76}$$

and

$$AF(\theta, \phi) = \sum_{n=1}^{N} I_n e^{j\beta R[\sin\theta \cos(\phi - \phi_n) - \sin\theta_0 \cos(\phi_0 - \phi_n)]} \tag{2-77}$$

It is noticed that circular arrays can be expressed in terms of uniformly excited and distributed ones by using a linear transformation on the excitation of elements. In this case, the pattern is given in terms of Bessel functions [19].

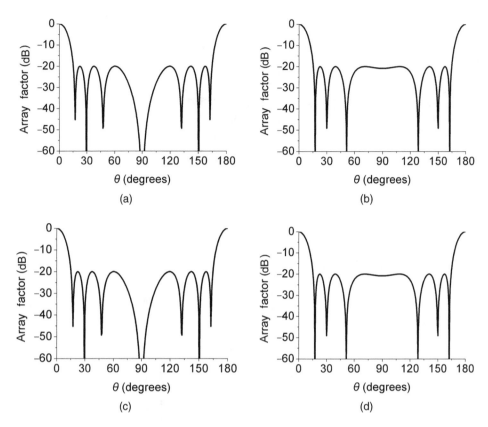

Figure 2.26   Planar patterns of a planar Chebyshev array for ϕ equal to (a) $0°$, (b) $30°$, (c) $90°$ and (d) $120°$

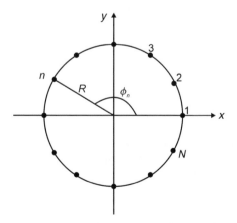

Figure 2.27   A circular array with N isotropic elements

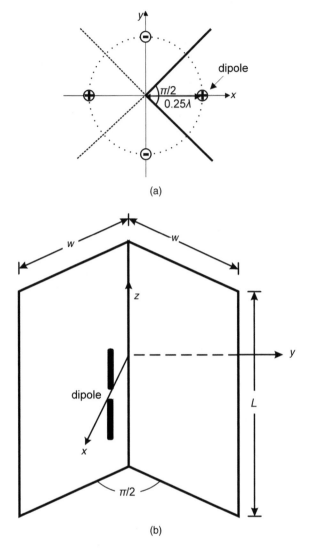

Figure 2.28   A dipole in front of a corner reflector with angle $\pi/2$, (a) Dipole and its images and (b) Perspective view

A dipole positioned at the bisector of a corner reflector with angle $\omega_N = \pi/N$ creates a circular array with $2N-1$ images (see Fig. 2.28). Figure 2.29 presents the corresponding H-plane pattern. It is noticed that, theoretically, the pattern outside the reflectors does not exist. In practice, this is not true because the sizes of the reflector's planes are finite and substantial diffraction on its edges occurs [19, 20].

An elliptical array, with its elements in an elliptical ring, produces a radiation pattern, which, in a similar manner, can be expressed in terms of Mathieu functions. The numerical computation appears to be difficult and a linear double transformation on the geometry and the excitation proposed in [21] can alleviate the problem.

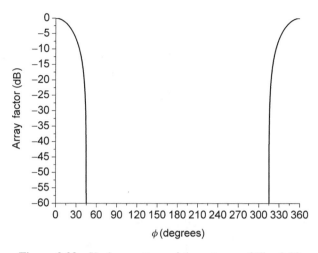

Figure 2.29   H-plane pattern of the antenna of Fig. 2.28

2.6.3 Ring Arrays

M circular arrays in concentric circular rings produce an array factor of the form

$$AF(\theta, \phi) = \sum_{m=1}^{M} \sum_{n=1}^{N} I_{mn} e^{j[\beta R_m \sin\theta \cos(\phi - \phi_{mn}) + \alpha_{mn}]} \tag{2-78}$$

$I_{mm} e^{j\alpha_{mn}}$ is the excitation of the nth element of the mth ring.

Ring arrays as well as circular ones are used to satisfy angular symmetry in two-dimensional operations [22].

A corner reflector with a linear array positioned in front of it (see Fig. 2.30) creates concentric rings. For a two-dipole uniform linear array, the pattern is presented in Fig. 2.31.

Concentric rings have been studied by using optimization methods. In [23], a series representation and a Monte Carlo simulation have been given for a concentric array design.

2.7 3-D Arrays

N elements positioned in 3-D space constitute a three-dimensional array. The array factor is

$$AF(\theta, \phi) = \sum_{n=1}^{N} I_n e^{j\{\alpha_n + \beta r_n [\sin\theta \sin\theta_n \cos(\phi - \phi_n) + \cos\theta \cos\theta_n]\}} \tag{2-79}$$

(r_n, θ_n, ϕ_n) are the spherical coordinates of the nth element and $I_n e^{j\alpha_n}$ is the corresponding excitation. A special case of a 3-D array is the cylindrical array. Parallel circular arrays with their centres on the same axis constitute a cylindrical array. The array factor is simplified to

$$AF(\theta, \phi) = \sum_{m=1}^{M} \sum_{l=1}^{L} I_{ml} e^{j[\alpha_{ml} + \beta R_0 \sin\theta \cos(\phi - \phi_{ml}) + \beta z_m \cos\theta]} \tag{2-80}$$

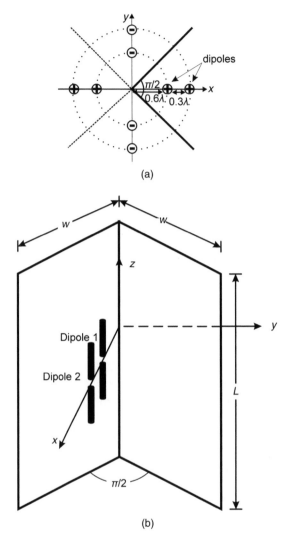

Figure 2.30 A linear array of dipoles in front of a $\pi/2$ corner reflector, (a) Two dipoles and their images and (b) Perspective view

$I_{ml}e^{j\alpha_{ml}}$ is the excitation of the lth element of the mth circular array.

R_0 is the radius of the circular arrays and z_m is the position of the mth array on the z-axis.

A cylindrical array can be made from an array of collinear dipoles in front of a corner reflector (Fig. 2.32).

To have a maximum at $\theta = \pi/2$, $\phi = \pi/4$, the array should be uniform. If there are additional constraints on the SLL, then the excitations should be non-uniform. Figure 2.33 shows the E and H-plane patterns of a Chebyshev array with $N = 9$ collinear dipoles in front of a $\pi/2$ corner reflector. $d = 0.7\lambda$, SLL ≤ -20 dB and maximum occurs at $\theta = \pi/2$, $\phi = \pi/4$.

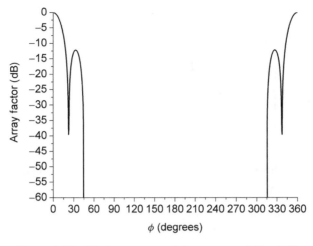

Figure 2.31   H-plane pattern of the antenna of Fig. 2.30

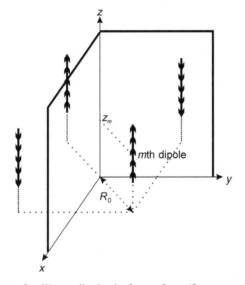

Figure 2.32   An array of collinear dipoles in front of a $\pi/2$ corner reflector and its images

2.8 Conformal Arrays

Arrays with requirements in conformity with a shaped surface are known as 'conformal'. Conformal arrays are used in mobile platforms for aerodynamic reasons. Also for specified angles of coverage, arrays can be conformal to stationary shaped surfaces.

The analysis of conformal arrays differs from that of planar and 3-D ones. The pattern of the array cannot be given by multiplying the element pattern and the array factor. They are not separable and, of course, the pattern is not a simple polynomial. Moreover, the illumination around the radiating surface as well as the polarization and the pattern of each radiating element must be separately taken into account.

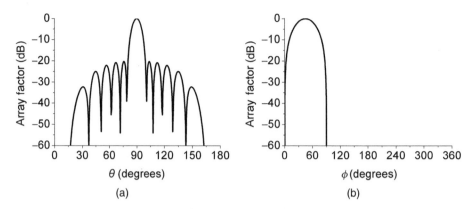

Figure 2.33 Patterns of a Chebyshev array with $N = 9$ collinear dipoles in front of a $\pi/2$ corner reflector and maximum at $\theta = \pi/2$, $\phi = \pi/4$ (a) E-plane pattern, (b) H-plane pattern

Practical communication and surveillance systems with scanning requirements use conformal arrays of cylindrical shape. In this case, one part of the array is illuminated. For the commutation of a given illumination region around the cylinder, one can use either mechanical rotation or switch networks, lens scanning, matrix beam formers or digital beam formers [24]. If the array compared to the radius R occupies a small sector and the radius is large ($R \gg \lambda$), then the element pattern approximates that of a planar array. If the sector is large compared to the radius R, or if the element is in the illuminated region of an array fully wrapped around the cylinder, then the pattern must be carefully calculated and is much more different than that of a planar array.

Other conformal arrays, such as the conical ones, are mainly used for missiles and aircrafts [25, 26]. The design follows that of cylindrical arrays. Grouping them in sub-arrays usually excites spherical, hemispherical and conical arrays. The problem of conformal arrays will be analysed in one of the next chapters, where the mutual coupling between the elements and the vector array factor will be taken into account.

References

[1] S.A. Schelkunoff and H.T. Friis, *Antenna Theory and Practice*, John Wiley & Sons, New York, 1952.

[2] C.A. Balanis, *Antenna Theory, Analysis and Design*, 3rd ed., John Wiley & Sons, New York, 2005.

[3] M. Abramowitz and L. Stegun, *Handbook of Mathematical Functions*, Dover Publications, New York, 1970.

[4] C.L. Dolph "A Current Distribution for Broadside Arrays Which Optimizes the Relationship between Beam Width and Side-Lobe Level," *Proc. IRE*, Vol. **34**, pp. 335–338, 1946.

[5] H.J. Riblet, "Discussion on a Current Distribution for Broadside Arrays which Optimizes the Relationship between Beam Width and Side-Lobe Level," *Proc. IRE*, Vol. **35**, pp. 489–492, 1947.

[6] G. Miaris, M. Chryssomalis, E. Vafiadis and J.N. Sahalos, "A Unified Formulation for Chebyshev and Legendre Superdirective End-fire Array Design," *Arch Elektrotech.*, Vol. **78**, No. 4, pp. 271–280, 1995.

[7] M. Dawoud and M.A. Hassan, "Design of Superdirective Endfire Antenna Arrays," *IEEE Trans. Antennas Propag.*, Vol. **37**, pp. 796–800, June 1989.

[8] R.S. Elliot, *Antenna Theory and Design*, Prentice Hall, Englewood Cliffs, NJ, 1981.

[9] R.C. Hansen (Ed.), *Microwave Scanning Antennas*, Vol. I, 1964, Vols. II & III, Academic Press, New York, 1966. (Peninsula Publishing 1985).

[10] A. Papoulis, *Signal Analysis*, McGraw-Hill, New York, 1984.

[11] R.S. Elliot, "On Discretizing Continuous Aperture Distributions," *IEEE Trans. Antennas Propag.*, Vol. **AP-25**, No. 5, pp. 617–621, 1977.

[12] T.T. Taylor, "Design of Line Source Antennas for Narrow Beamwidth and Low Sidelobes," *IEEE Trans. Antennas Propag.*, Vol. **AP-3**, pp. 16–28, 1955.

[13] A.T. Villeneuve, "Taylor Patterns for Discrete Arrays," *IEEE Trans. Antennas Propag.*, Vol. **AP-32**, No. 10, pp. 1089–1093, 1984.

[14] R.C. Hansen, "Linear Arrays," in *Handbook of Antenna Design*, Vol. 2, Chap. 9, A. Rudge (Ed.), Peter Peregrinus, London, 1983.

[15] E.T. Bayliss, "Design of Monopulse Antenna Difference Patterns with Low Sidelobes," *Bell Syst. Tech. J.*, Vol. **47**, pp. 623–650, 1968.

[16] R.C. Mailloux, *Phased Array Antenna Handbook*, Artech House, Norwood, MA, 1994.

[17] Ye.V. Baklanov, "Chebyshev Distribution of Currents for a Plane Array of Radiators," *Radio Eng. Electron. Phys.*, Vol. **11**, pp. 640–642, April 1966.

[18] F.I. Tseng and D.K. Cheng, "Optimum Scannable Planar Arrays with an Invariant Side Lobe Level," *Proc. IEEE*, Vol. **56**, pp. 1771–1778, 1968.

[19] Y.T. Lo, "Array Theory," *Antenna Handbook*, Chap. 11, Van Nostrad Reinhold, New York, 1988.

[20] C.E. Ryan Jr and L. Peters Jr. "Evaluation of Edge Diffracted Fields Including Equivalent Currents for Caustic Regions," *IEEE Trans. Antennas Propag.*, Vol. **AP-17**, pp. 292–299, 1969.

[21] Y.T. Lo and H.C. Hsuan, "An Equivalence Theory between Elliptical and Circular Arrays," *IEEE Trans. Antennas Propag.*, Vol. **AP-13**, pp. 247–256, 1965.

[22] M.T. Ma, *Theory and Applications of Antenna Arrays*, Wiley-Interscience, New York, 1974.

[23] W.F. Richards and Y.T. Lo, "Antenna Pattern Synthesis Based on Optimization in a Probabilistic Sense," *IEEE Trans. Antennas Propag.*, Vol. **AP-23**, pp. 165–172, 1975.

[24] W.H. Kummer (G. Ed.), "Special Issue on Conformal Arrays", *IEEE Trans. Antennas Propag.*, Vol. **AP-22**, No. 1, 1974.

[25] T.N. Kaifas and J.N. Sahalos, "On the Design of Cylindrical Conformal Microstrip Arrays," *Proc. of 2nd International Symposium of BSUAE*, p. 18, Xanthi (Greece), 2000.

[26] T.N. Kaifas, Ch. Koukourlis and J.N. Sahalos, "On the Beam Steering of Cylindrical Conformal Microstrip Antenna Arrays," *AP 2000 Millennium Conference on Antennas and Propagation*, Davos (Switzerland), April 2000.

3

Pattern Synthesis for Arrays

3.1 Introduction

One of the main advantages of arrays is that they can realize adequate approximations of certain desired radiation patterns. Several techniques have been given in the past for synthesizing array factors. Most of them have to do with the synthesis of narrow-beam and/or low side-lobe patterns. In the literature [1–4], one can find elegant case studies useful in wireless applications.

The synthesis is based basically on the antenna engineer's experience and ingenuity, who aims at realizable solutions and approaches the desired properties of the array.

In the following paragraphs, several synthesis procedures will be briefly described.

3.2 Uniform Linear Array Synthesis

Uniform arrays can be used for SLL ≥ -13.2 dB. This is a limit coming from the SLL of a uniform line source. A linear scanning array with maximum at $\theta = \theta_0$ and half power beam width θ_H must have (see Table 2.1)

$$\theta_0 > \cos^{-1}\left(\cos^2 \frac{\theta_H}{2}\right) \tag{3-1}$$

and

$$N\frac{d}{\lambda} = 0.4428 \left[\frac{2(1 + \cos \theta_H)}{\sin^4 \theta_0 - (\cos \theta_H - \cos^2 \theta_0)^2}\right]^{1/2} \tag{3-2}$$

For example,

$$\left.\begin{array}{l} \text{for } \theta_H = 10° \text{ and } \theta_0 = 90° \Rightarrow N\dfrac{d}{\lambda} = 5.08 \quad \text{(broadside array)} \\[2mm] \text{for } \theta_H = 10° \text{ and } \theta_0 = 30° \Rightarrow N\dfrac{d}{\lambda} = 10.28 \quad \text{(intermediate array)} \end{array}\right\} \tag{3-3}$$

Orthogonal Methods for Array Synthesis: Theory and the ORAMA Computer Tool   John N. Sahalos
© 2006 John Wiley & Sons, Ltd

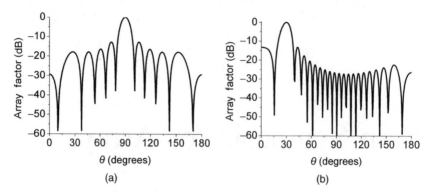

Figure 3.1   Patterns of uniform arrays with HPBW $= 10°$ (a) Broadside with $N = 8$, (b) Intermediate with $N = 23$

Figures 3.1a and b show the patterns of an $N = 8$ element broadside array ($d/\lambda = 0.635$) and an $N = 23$ element ($d/\lambda = 0.447$) intermediate array with the same HPBW given in Equation (3-3).

For an end-fire array, where $\theta_0 = 0°$, Table 2.1 gives

$$N\frac{d}{\lambda} = \frac{0.4428}{1 - \cos\dfrac{\theta_H}{2}} \quad \text{(Ordinary end-fire)} \tag{3-4}$$

$$N\frac{d}{\lambda} = \frac{0.1398}{1 - \cos\dfrac{\theta_H}{2}} \quad \text{(Hansen-Woodyard)} \tag{3-5}$$

Uniform arrays are useful in practice. A linear array for mobile communications with maximum at $\theta_0 = 98°$ and $\theta_H = 10°$ must have

$$N\frac{d}{\lambda} = 5.162 \tag{3-6}$$

Seven $\lambda/2$ collinear dipoles with phase difference of $37°$ in $d/\lambda = 0.737$ can be used to implement the antenna array. The E-plane pattern (see Fig. 3.2) is derived by the multiplication of the array factor and the element pattern.

3.3 Chebyshev Array Synthesis

Arrays with constraints on the SLL are useful in communications and radar systems. The Dolph method can be used, in general, for $d/\lambda \geq 0.5$. To avoid grating lobes, we must have $d \leq d_{max}$, where

$$\frac{d_{max}}{\lambda} = 1 - \frac{\cos^{-1}(1/x_0)}{\pi} \tag{3-7}$$

x_0 is the distance where the maximum of the array occurs.

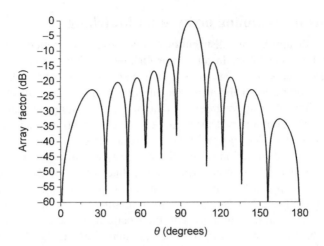

Figure 3.2 E-plane pattern of a uniform array of seven collinear dipoles with $d/\lambda = 0.737$, $\theta_0 = 98°$ and $\theta_H = 10°$

For $d/\lambda < 0.5$, the Riblet method for odd numbers of elements can be used. For the Dolph array, the HPBW has been related to that of the uniform one (HPBW$_u$) of the same length. The so-called broadening factor f was found to be [1]

$$f = \frac{(\text{HPBW})}{(\text{HPBW}_u)} = 1 + 0.632 \left[\frac{2}{R} \cosh\sqrt{(\cosh^{-1}R)^2 - \pi^2} \right]^2 \tag{3-8}$$

f is valid in the range of -60 dB $\leq$ SLL ≤ -20 dB and for scanning near broadside. With the help of f, directivity D is estimated by

$$D = \frac{2R^2}{1 + (R^2 - 1)f \dfrac{\lambda}{Nd}} \tag{3-9}$$

For the Riblet case, the HPBW is given approximately by [1, 2]:

$$\text{HPBW} \cong 10.3° \frac{\lambda}{Nd} \sqrt{s + 4.52} \csc\theta_0 \tag{3-10}$$

where s is the SLL in dB and θ_0 is the scan angle.
Also, directivity D is

$$D = \frac{2R^2}{1 + R^2 \dfrac{\lambda}{Nd} \sqrt{\dfrac{\ln(2R)}{\pi}} \cdot \sin\left(\beta\dfrac{d}{2}\right)} \tag{3-11}$$

By removing $\sin\left(\beta\dfrac{d}{2}\right)$ from Equation (3-11), the above expression is valid also for arrays with element spacing greater than $\lambda/2$ but less than that of the appearance of grating lobes.

3.4 Synthesis by Sampling or by Root Matching

Line sources with appropriate distributions create excellent patterns with the desired HPBW and/or low side lobes. In an earlier chapter, we have seen that arrays coming from sampling of line sources can give similar patterns. It is noticed that, for large element spacing, the patterns of the array and the line source do not match well. A useful and elegant method of array synthesis is that of root matching. In particular, the nulls of the pattern of the continuous distribution are matched with the pattern of the discrete array. If the final pattern does not yield the desired accuracy, a perturbation technique [3] is applied to vary the distribution of the discrete array and to improve accuracy.

An array of fixed length is transformed into a line source of the same length as the number of elements approaches infinity. The array factor reduces to the space factor (SF) that comes from the finite 1-D Fourier transform relating the far field to the excitation. In Chapter 2, several useful distributions with their pattern characteristics have been given. Among them, the Chebyshev error, the one-parameter Taylor, the Bayliss, the iterative Taylor, the iterative Bayliss and the Orhard are the most interesting. In this paragraph, we do not intend to repeat their details. The reader can go back to the previous chapter for more information.

3.5 Synthesis by Fourier Transform

A linear array with uniformly spaced elements and non-uniform excitation has an array factor of the form

$$AF(\psi) = \sum_{n=1}^{N} I_n e^{jn\psi} \tag{3-12}$$

A desired array factor $AF_d(\psi)$ can be expanded in a Fourier series with infinite terms ($N \to \infty$) of the form as in Equation (3-12). The first N coefficients of the expansion of the two array factors are equated to approximate the desired pattern. The coefficients are found by using the orthogonal character of the expansion functions:

$$I_n = \frac{1}{2\pi} \int_{-\pi}^{\pi} AF_d(\psi) e^{-jn\psi} \, d\psi \tag{3-13}$$

The Fourier method is adequate for spacing $d = 0.5\lambda$. For $d > 0.5\lambda$, it fails, while for $d < 0.5\lambda$ and a sufficient number of elements, the pattern matches the desired one better. The radiation pattern of an array with $N = 17$ elements and $d = 0.5\lambda$ is presented in Fig. 3.3. The desired pattern is a pulse between $80° \leq \theta \leq 100°$. The pattern for the same array with $d = 0.35\lambda$ and $d = 0.75\lambda$ is shown in Figs. 3.4a and b.

3.6 The Woodward – Lawson (WL) Method

A uniform linear array with N elements has an array factor of the form $\frac{\sin(N\psi/2)}{N\sin(\psi/2)}$. This factor is the narrowest that can be achieved with an array. The pattern of a uniform array is an alternative and a useful tool for the synthesis of arrays because it can be regarded as a member of an orthogonal set of beams. By superimposing groups of beams [1, 2],

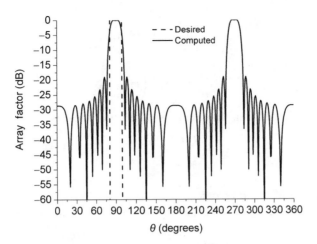

Figure 3.3 Pattern of a linear array with $N = 17$ elements, $d/\lambda = 0.5$ with a pulse at $80° \leq \theta \leq 100°$ as a desired pattern

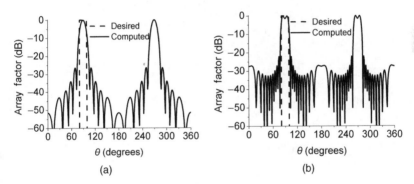

Figure 3.4 Pattern of a linear array with $N = 17$ elements, with a pulse at $80° \leq \theta \leq 100°$ as a desired pattern (a) $d/\lambda = 0.35$ and (b) $d/\lambda = 0.75$

we can synthesize a desired pattern. A uniform array with N elements in spacing d/λ produces a beam pattern of the form

$$f_m(\theta) = b_m \frac{\sin(N\psi_m/2)}{N\sin(\psi_m/2)} \tag{3-14}$$

where

$$\psi_m = \beta d(\cos\theta - \cos\theta_m) \tag{3-15}$$

A desired factor is assumed to be the superimposition of the terms given in Equation (3-14). Thus, $AF(\theta)$ is

$$AF(\theta) = \sum_{m=-M}^{M} b_m \frac{\sin(N\psi_m/2)}{N\sin(\psi_m/2)} \tag{3-16}$$

where

$$b_m = AF(\theta_m) \tag{3-17}$$

and

$$\theta_m = \cos^{-1}\left(m\frac{\lambda}{Nd}\right) \tag{3-18}$$

Due to the superimposition, the excitation of each element is

$$I_n = \frac{1}{N}\sum_{m=-M}^{M} AF(\theta_m)e^{-j\beta d_n \cos\theta_m} \tag{3-19}$$

d_n is the position of the nth element.

For a line source, we again can superimpose groups of beams of the form [1]

$$f_m(\theta) = b_m\frac{\sin[\beta\,{}^L/_2(\cos\theta - \cos\theta_m)]}{\beta\,{}^L/_2(\cos\theta - \cos\theta_m)} \tag{3-20}$$

The SF is

$$SF(\theta) = \sum_{m=-M}^{M} b_m\frac{\sin[\beta\,{}^L/_2(\cos\theta - \cos\theta_m)]}{\beta\,{}^L/_2(\cos\theta - \cos\theta_m)} \tag{3-21}$$

where

$$b_m = SF(\theta_m) \tag{3-22}$$

and

$$\theta_m = \cos^{-1}\left(m\frac{\lambda}{L}\right) \tag{3-23}$$

Finally, the excitation distribution is

$$I(Z') = \sum_{m=-M}^{M} b_m e^{-j\beta z' \cos\theta_m} \tag{3-24}$$

3.7 Array Synthesis as an Optimization Problem

In general, antenna array synthesis can be characterized as a non-linear optimization problem. In this problem, an appropriate real function, which takes an optimum value at the reached properties of the desired array, is constructed. Several functions can be used to fit specific antenna properties. Such properties are [4]:

1. The radiation pattern at specific frequency/ies.
2. The antenna impedance at specific frequency/ies.
3. An antenna index (gain, directivity etc.) under or without constraint on another index.
4. The antenna impedance and/or radiation pattern at specific frequency ranges.
5. The antenna coupling.

The parameters that optimize a desired function may determine one or more characteristics of the array. Such characteristics are the excitation, the shape, the size, the loadings and the current distribution of the antenna elements. Any variation in the desired parameters of the array requires completely new solutions.

Antenna optimization methods can be divided into two categories. The first one takes the values of the optimization function itself into account. The second one looks at the gradient of the above function. The optimization function is not an explicit function in some cases. It can be described at M points and it is simply computed numerically. If the number of points is equal to the unknown $(M = N)$, then it is possible to have one solution. In $M < N$, the problem is underdetermined and more than one solutions exist. Finally, if $M > N$, the problem is over determined and a least squares solution exists.

It is noticed that elegant procedures based on random search are also available [4]. The above procedures make use of random number generators by which the successive points are determined. Finally, the Simulated Annealing (SA) and the Genetic Algorithms (GA) are two global optimizer methods [5–7].

3.7.1 Optimization of an Array Index

An antenna array index, I, such as directivity, gain and so on, can be written as

$$I = \frac{[\tilde{a}]^*[A][a]}{[\tilde{a}]^*[M][a]} \tag{3-25}$$

where $[\tilde{a}]^* = [a_1{}^* a_2{}^* \ldots a_N{}^*]$ is the conjugate transpose of $[a]$. By $[a]$, one can represent the current or the voltage excitation vector of the array. $[A]$ and $[M]$ with

$$\left.\begin{array}{l} [A] = [\alpha_{ij}] = [\tilde{B}]^*[B] \\ [M] = [m_{ij}] \end{array}\right\} \tag{3-26}$$

are both Hermitian $N \times N$ square matrices. Also, $[M]$ is positive definite and $[B]$ is a row vector [8].

An index I of the form in Equation (3-25) will be optimized under the constraint that another index I_1 is

$$I_1 = \frac{[\tilde{a}]^*[M_2][a]}{[\tilde{a}]^*[M_3][a]} = \gamma \tag{3-27}$$

According to [8–11], a solution can be found by using the Lagrange multiplier and by setting the quantity L,

$$L = \frac{[\tilde{a}]^*[A][a]}{[\tilde{a}]^*[M][a]} + \lambda \left\{ \frac{[\tilde{a}]^*[M_2][a]}{[\tilde{a}]^*[M_3][a]} - \gamma \right\} \tag{3-28}$$

to be stationary with respect to $[a]$ and λ.

By zeroing the first variation of L, we have

$$[a] = q[K]^{-1}[\tilde{B}]^* \tag{3-29}$$

q is a constant and

$$[K] = [M] + p\{y[M_3] - [M_2]\} \tag{3-30}$$

p is found from Equation (3-27), which is modified to an eigenvalue equation [10–11]. If there is no constraint on I_1, $p = 0$.

In the optimization procedure, pattern values and/or index constraints can also be combined. A special case of optimization is the one where the constraints have to do with specific pattern nulls. The problem is solved in two steps [12–13]. In the first step, a set of M elements (M is the number of nulls) of [a] is expressed as a linear relation of the rest $N - M$. In the second step, the Langrangian of (3–28) is modified into an expression with vectors that contain the $N - M$ independent elements. By following a similar procedure as above, the vectors are derived.

3.7.2 Optimization by Simplex and Gradient Methods

Simplex and gradient [14–16] are both local optimizer methods. A simplex is an n-dimensional space, which constitutes a polyhedron with $n + 1$. Actually, it is the generalization of the tetrahedron in 3-D. The simplex method computes the optimization function at the vertices of a simplex. On the basis of the values of the above function, a new smaller simplex is chosen. The process is repeated until a sufficiently small simplex locates an optimum with desired accuracy. The optimum depends on the initial simplex [14]. It is noticed that a simplex cannot always find a global optimum.

Gradient methods are known as 'steepest-descent methods'. A starting point is chosen and the direction in a multidimensional space where the optimization function decreases most rapidly is found. By adopting a new point in that direction at a desired distance from the previous one and by repeating the process, a minimum of the optimization function is achieved. A gradient method is not faster than a simplex method. Also, it is difficult to judge if the achieved minimum is a global or a local one. From an antenna-engineering point of view, we are usually interested in a practically suitable solution and not necessarily in the global optimum. The following example illustrates the optimization of the radiation pattern and the antenna impedance of a four-element Yagi–Uda array. For the pattern, the antenna gain obtained is larger than the prescribed value. For the impedance, the mutual coupling is taken into account. Initial values for the optimization parameters are used based on experience. After 12 simplex iterations, an antenna is obtained with (see Fig. 3.5) $L_R = 0.48\lambda$, $L_F = 0.457\lambda$, $L_D = 0.45\lambda$, $d = d_r = 0.30\lambda$.

It is found that $D = 11$ dB, $Z_{in} = 38\ \Omega$ and HPBW $= 56.2°$. Figure 3.6 shows the pattern of the antenna. A similar optimization can be applied to other arrays, [17].

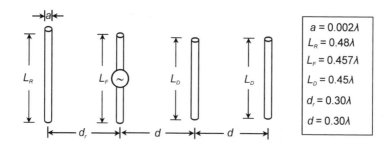

Figure 3.5   Geometry of a four-element Yagi–Uda array

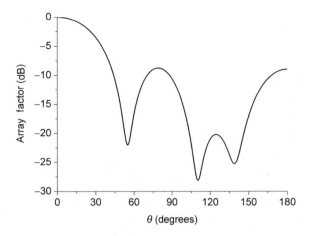

Figure 3.6   Pattern of the Yagi–Uda array given in Fig. 3.5

3.7.3 Optimization by Simulated Annealing Method

Simulated annealing (SA) is a method that combines local search with Monte Carlo techniques by analogy with cooling processes in thermodynamics. SA [18] refers to a process of revealing the low temperature state of some material. At high temperatures, the molecules of a liquid move freely with respect to one another. If the liquid is slowly cooled, its thermal mobility is lost. The atoms are able to line up in a crystal. The crystal represents the minimum energy state for the system. The time spent at each temperature must be sufficiently long to allow a thermal equilibrium to be realized. If the system is cooled quickly, it does not reach the minimum energy state but one having higher energy.

It was mentioned previously that in SA we simulate the annealing process by a Monte Carlo method. In this case, the global minimum of the objective function represents the low energy configuration, [6, 19]. A SA process can include parallelization techniques with parallel processing of multiple CPUs, [7]. Various combinatorial optimization problems have used SA.

An example of an array of eight $\lambda/2$ collinear dipoles is presented. The initial array is a uniform array with $d/\lambda = 0.93$ and phase shift $\alpha = 52°$. The main-beam maximum is at $\theta = 99°$. It is observed that, in the area $\theta \leq 90°$, an SLL ≥ -10 dB occurs. This, in practice, is not desirable. Mobile and radio stations aim at lower levels upwards of the horizon. By using the appropriate cost function with the geometry constraints for SLL ≤ -18 dB in $\theta \leq 90°$, a new array is found. Table 3.1 and Fig. 3.7 present the array and the pattern. Different applications for wire antenna arrays as well as for slot arrays can be found in [20–21].

Table 3.1   Position and phase of an optimized linear array

Dipole number	1	2	3	4	5	6	7	8
Dipole position (λ)	0	0.70	1.53	2.37	3.23	4.05	4.92	5.62
Dipole phase (deg)	0	73	116	155	−172	−130	−93	−19

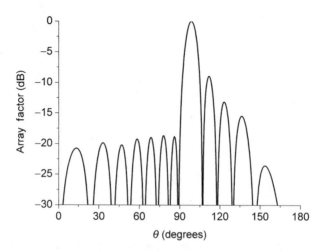

Figure 3.7 E-plane pattern of $N = 8$ collinear dipoles with SLL ≤ -18 dB for $\theta \leq 90°$ and max at $\theta = 99°$

3.7.4 Optimization by Genetic Algorithms (GAs)

GAs are global optimizers based on natural selection and genetics mechanisms. GAs [5] follow two main principles: the ability to encode complex structures and the use of simple transformations to improve such structures. GAs have become attractive for a wide range of problems in electromagnetics [5]. They have the advantage of quick and easy programming and implementation over other methods. Also, they are suitable for constrained optimization. GAs are based on Darwin's evolutionary principle [22]: survival of the fittest. The basic idea is an analogy between an individual and a solution on the one hand and between an environment and a given problem on the other. The function to be minimized or maximized represents the fitness. This is computed for a given individual and determines how he/she fits or, in other words, how good the solution to a given problem is.

Many categories of GAs have been designed. The simple GA, as it is presented in [23], has non-overlapping populations. Popular in electromagnetics are the Steady State GAs. They contain overlapping populations where the best individuals survive to the next generation. Another approach is the Deme GA [24] that combines parallel evolving populations with migration.

Since the early 1990s, GAs have been successfully applied in several engineering problems [25–28]. Among them, the thinned array problems are of interest. A thinned array is a subset of aperiodic arrays. Thinning an array means turning off some elements in a uniformly spaced or periodic array. The off elements remain in the array, so that the mutual coupling remains the same for the interior elements.

A thinned array offers essentially the same beam width with less directivity and fewer elements than a uniform array of the same size [5, 29, 30]. The most realistic applications of GAs to array thinning have to do with optimizing the SLL of a large number of elements.

3.7.5 Space and Time Optimization/Smart Antennas

The limited availability of spectrum, the increasing demand for more users and the high bit rate services has made wireless operators to explore new options. Antenna arrays combined with signal processing in space and real time, known as 'smart antennas', provide new technological options. Low cost and fast digital processors nowadays can ensure the implementation of smart antennas. Smart antennas start to appear with great success in cellular and satellite mobile communication systems. They improve their performance by increasing both the spectrum efficiency and the channel capacity. Smart antennas extend the range of coverage by multiple beam steering and electronic compensation of the distortion. They can also reduce propagation problems such as multipath fading, co-channel interference and delay spread. Their main advantage is their capability of providing a certain channel in a certain direction. This is known as 'Spatial Division Multiple Access' (SDMA). The Spatial performs in a different way than the frequency division multiple access (FDMA), the time division multiple access (TDMA) and the code division multiple access (CDMA) [31].

Smart antennas are known as 'adaptive arrays', 'intelligent antennas', 'spatial processing', 'digital beam forming antennas' and so on, [32–35]. They direct their main-beam maximum to the user while the pattern nulls are in the direction of possible interference, [36].

Three main types of beam patterns are available. These are the switched-beam, the dynamic phased arrays and the adaptive antennas. The switched-beam divides the communication sector into smaller pre-defined sectors. Each small sector contains a predetermined fixed beam pattern. Dynamic phased arrays make use of the direction of arrival (DoA) information from the desired user and steer the beam towards the user. Adaptive antennas dynamically alter the patterns to optimize communications performance. They utilize sophisticated signal processing algorithms [36, 37], which update the beam patterns based on changes in both the desired and the interfering signal directions.

Adaptive array theory is based on the previously mentioned optimization methods and on real-time response in a transient environment.

In the literature, one can find a lot of special issues, books and specialized research papers in the area of smart antennas [38–43].

Smart antenna arrays are analysed for different network topologies and mobility scenarios. The arrays are realized with appropriate feed networks and with algorithms that derive the DoA and the fast beam forming.

Smart antennas have made tremendous progress in this decade. It is believed that the development of new array processing algorithms and the efforts in cost reduction will shortly make them preponderant in wireless systems.

3.8 Synthesis by Convolution of Linear, Planar and 3-D Arrays

Planar arrays, as they were analysed in Chapter 2, can be combined with two linear arrays through pattern multiplication. Also, it was mentioned that, for equal side lobes in all ϕ cuts, a single Chebyshev pattern can be realized. Circular or elliptical or ring arrays can be synthesized by sampling the corresponding continuous distributions. The number of samples must be adequate in order to produce a similar pattern.

Another interesting method of synthesis of planar arrays can come from the convolution procedure. In this case, the desired pattern is a multiplication of two or more simpler patterns. It is known that this pattern is the Fourier transform of source distribution in space. If a pattern is the product of two simpler patterns, then the corresponding distribution is the convolution of the two distributions, [44, 45].

Let us consider two arrays described as the distribution of weighted impulse functions:

$$W_1(r) = \sum_{i=1}^{N_1} w_{1i}\delta(r - r_{1i}) \tag{3-31}$$

$$W_2(r) = \sum_{j=1}^{N_2} w_{2j}\delta(r - r_{2j}) \tag{3-32}$$

w_{1i}, w_{2j} are the excitation coefficients, and r_{1i}, r_{2j} is the location of the ith and jth element of the arrays.

The array with a pattern equal to the product of the pattern of these two arrays is found by convolving the two distributions. Thus, we have

$$W_1(r) * W_2(r) = \sum_{i=1}^{N_1}\sum_{j=1}^{N_2} w_{1i} \cdot w_{2j}\delta(r - r_{1i} - r_{2j}) \tag{3-33}$$

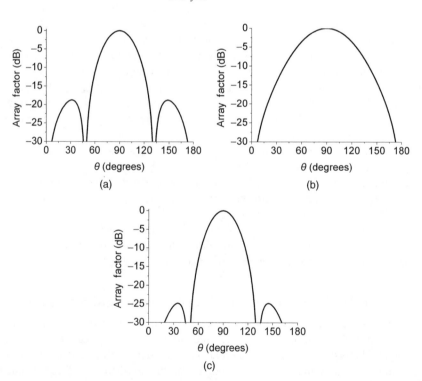

Figure 3.8   Radiation patterns of (a) A linear array with $N = 3$ and $d = 0.5\lambda$, (b) A linear array with $N = 2$ and $d = 0.4\lambda$ and (c) The convolution of (a) and (b)

In general, the convolution is performed on three axes and the distribution can be a 3-D array. Special cases of linear and planar arrays occur when the two simpler arrays are on the same or on normal axes respectively.

Two linear arrays on the same axis produce a new array whose excitation is the product of the excitations $(w_{1j} \cdot w_{2j})$ and its position, the summary of the positions $(r_{1i} + r_{2j})$. Consider an example of convolution of two uniform linear arrays with $N = 3$ and spacing 0.5λ for the first, and $N = 2$ with spacing 0.4λ for the second. The resulting array is a linear one with equal excitations, $N = 6$ elements and positions of the elements at $0\lambda, 0.4\lambda, 0.5\lambda, 0.9\lambda, 1.0\lambda$ and 1.4λ. Figrue 3.8 presents first the pattern of the arrays separately and then the final pattern.

If the spacing of both arrays is the same, let us say 0.5λ, then the number of the elements of the final array is $N = 4$. This happens because some of the elements of the convolution are in the same place. In this case, the excitation of the final array is 1, 2, 2, 1.

Consider two linear arrays, one with $N = 3$ and spacing 0.5λ on the x-axis and the other with $N = 2$ and spacing 0.5λ on the y-axis. The coordinates of the six elements of the final array will be $(0.0\lambda, 0.0\lambda)$, $(0.0\lambda, 0.5\lambda)$, $(0.0\lambda, 1.0\lambda)$, $(0.5\lambda, 0.0\lambda)$, $(0.5\lambda, 0.5\lambda)$, $(0.5\lambda, 0.1\lambda)$, and the excitation will be uniform.

Rhombic arrays with a building block of a simple rhombic array can be designed. Also, the convolution of a rhombic array with a linear one can give the desired nulls of the final pattern. One can find useful examples in the literature [2, 45].

References

[1] C.A. Balanis, *Antenna Theory, Analysis and Design*, 3rd ed., John Wiley & Sons Inc, New York, 2005.

[2] T. Milligan, *Modern Antenna Design*, McGraw-Hill, New York, 1985.

[3] R.S. Elliot, "On Discretizing Continuous Aperture Distributions," *IEEE Trans. Antennas Propag.*, Vol. **AP-25**, No. 5, pp. 617–621, 1977.

[4] B.D. Popovic, M.B. Dragovic and A.R. Djordjevic, *Analysis and Synthesis of Wire Antennas*, Research Studies Press, John Wiley & Sons, New York, 1982.

[5] Y. Rahmat-Samii and E. Michielssen, *Electromagnetic Optimization by Genetic Algorithms*, Wiley-Interscience, New York, 1999.

[6] A. Torn and A. Zilinskas, *Global Optimization, Lecture Notes in Computer Science*, Springer-Verlag, Berlin, 1987.

[7] R. Azencott, *Simulated Annealing Parallelization Techniques*, John Willey & Sons, New York, 1992.

[8] Y.T. Lo, S.W. Lee and Q.H. Lee, "Optimization of Directivity and SNR of an Arbitrary Antenna Array," *Proc. IEEE*, Vol. **54**, No. 8, pp. 1033–1045, 1966.

[9] R.F. Harrington, *Field Computations by Moment Method*, IEEE Press, New York, 1993.

[10] L.P. Winkler and M. Schwartz, "A Fast Numerical Method for Determining the Optimum SNR of an Array Subject to a Q Factor Constraint," *IEEE Trans. Antennas Propag.*, Vol. **AP-20**, No. 4, pp. 503–505, 1972.

[11] P. Zimourtopoulos and J.N. Sahalos, "On the Gain Maximization of the Dual Frequency and Direction Array Consisting of Wire Antennas," *IEEE Trans. Antennas Propag.*, Vol. **AP-33**, pp. 874–880. 1985.

[12] C.J. Drane, Jr and J.F. McLlvenna, "Gain Maximization and Controlled Null Placement Simultaneously Achieved in Aerial Array Patterns," *Physical Sciences Research Paper No. 383, AFCRL-69-0257*, LG Hanscom Field, Bedford, MA, June 1969.

[13] A.T. Adams and B.J. Strait, "Modern Analysis Methods for EMC," *1970 IEEE EMC Symposium Record*, pp. 383–393, July 1970.

[14] J.A. Nelder and R. Mead, "A Simplex Method for Function Minimization," *Comp. J.*, Vol. **7**, pp. 308–313. 1965.

[15] S.L.S. Jacoby, J.S. Kowalik and J.T. Pizzo, *Iterative Methods for Nonlinear Optimization Problems*, Prentice-Hall, Englewood Cliffs, NJ, 1972.

[16] P.R. Abdy and M.A.H. Dempster, *Introduction to Optimization Methods*, Chapman and Hall, London, 1974.

[17] M.T. Ma, *Theory and Applications of Antenna Arrays*, Wiley-Interscience, New York, 1974.

[18] N. Metropolis, A.W. Rosenbluth, A.H. Teller and E. Teller, "Equations of State Calculations by Fast Computing Machines," *J. Chem. Phys.*, Vol. **21**, pp. 1087–1091. 1953.

[19] W.H. Press, S.A. Teukolsky, W.T. Vetterling and B.P. Flannery, *Numerical Recipes in C*, 2nd ed., Cambridge University Press, Cambridge, 1992.

[20] Z. Zaharis, E. Vafiadis and J.N. Sahalos, "On the Design of a Dual-Band Base Station Wire Antenna", *IEEE Antennas Propag. Mag.*, Vol. **42**, No. 6, pp. 144–151, 2000.

[21] K. Kechagias, E. Vafiadis and J.N. Sahalos, "On the RLSA Antenna Optimum Design for DBS Reception", *IEEE Trans. Broadcast.*, Vol. **44**, No. 4, pp. 460–469, 1998.

[22] C. Darwin, *On the Origin of Species*, John Murray, London, 1859.

[23] D. Goldberg, *Genetic Algorithms in Search Optimization and Machine Learning*, Addison-Wesley, Reading, MA, 1989.

[24] M. Wall, G. Alib: *A C + + Library of Genetic Algorithm Components*, Version 2.4, Document Revision B, MIT, Boston, 1996.

[25] J.H. Holland, *Adaptation in Natural and Artificial Systems*, University of Michigan Press, Ann Arbor, 1975.

[26] M. Mitchell, *An Introduction to Genetic Algorithms*, 2nd edn., MIT Press, Boston, 1996.

[27] L. Chambers, *Practical Handbook of Genetic Algorithms*, Applications Volume I, CRC Press, New York, 1995.

[28] K.F. Man, K.S. Tang and S. Kwong, *Genetic Algorithms*, Springer-Verlag, London, 1999.

[29] R.C. Mailloux, *Phased Array Antenna Handbook*, Artech House, Norwood, MA, 1994.

[30] R.C. Hansen, *Phased Array Antennas*, Wiley-Interscience, New York, 1998.

[31] G. Maral and M. Bousquet, *Satellite Communications Systems*, 2nd edn., Wiley, New York, 1993.

[32] T.S. Rappaport (Ed.), *Smart Antennas*, IEEE Press, NJ, 1998.

[33] G.V. Tsoulos (Ed.), *Adaptive Antennas for Wireless Communications*, IEEE Press, N.J., 2001.

[34] M. Chryssomallis, "Smart Antennas", *IEEE Antennas Propag. Mag.*, Vol. **42**, No. 3, pp. 129–136, 2000.

[35] R. Janaswamy, *Radiowave Propagation and Smart Antennas for Wireless Communications*, Kluwer Academic Publishers, Boston, 2001.

[36] R.T. Compton Jr, *Adaptive Antennas*, Prentice-Hall, Englewood Cliffs, NJ, 1988.

[37] J.G. Proakis and D.G. Manolakis, *Digital Signal Processing*, 2nd edn., Mc Millan, New York, 1992.

[38] R. Schreiber, "Implementation of Adaptive Array Algorithms", *IEEE Trans. Acoustics, Speech Signal Process.*, Vol. **ASSP-34**, pp. 1038–1045. 1986.

[39] S. Choi and T.K. Sarkar, "Adaptive Antenna Array Utilizing the Conjugate Gradient Method for Multipath Mobile Communications", *Signal Process.*, Vol. **29**, pp. 319–333. 1992.

[40] A. El Zooghby, C.G. Christodoulou and M. Georgiopoulos, "Neural Network-based Adaptive Beamforming for One and Two Dimensional Antenna Arrays", *IEEE Trans. Antennas Propag.*, Vol. **AP-46**, pp. 1891–1893. 1998.

[41] Th.S. Rappaport (Ed.), *Smart Antennas: Adaptive Arrays, Algorithms and Wireless Position Location*, IEEE Press, NJ, 1998.

[42] L.C. Godara. "Application of Antenna Arrays to Mobile Communications, Part I: Performance Improvement, Feasibility and System Considerations", *Proc. IEEE*, Vol. **85**, No. 7, pp. 1031–1060, 1997.

[43] L.C. Godara, "Application of Antenna Arrays to Mobile Communications, Part II: Beam-Forming and Direction-of-Arrival Considerations", *Proc. IEEE*, Vol. **85**, No. 8: pp. 1195–1245, 1998.

[44] B.D. Steinberg, *Principles of Aperture & Array System Design*, Wiley, New York, 1976.

[45] S.R. Laxpati, "Planar Array Synthesis with Prescribed Pattern Nulls", *IEEE Trans. Antennas Propag.*, Vol. **AP-30**, No. 6, 1176–1183, November 1982.

4

The Orthogonal Methods

4.1 Introduction

In the early days of radio communications, the importance of the realization of appropriate efficient antennas was self-evident. During the 1920s, when short waves became popular, the need for antenna arrays was indisputable. It would be useful to go over some of the studies during the last six decades before we come to the discussion of orthogonal methods.

Wolff [1] was a pioneer of the synthesis of uniformly distributed linear arrays with circularly symmetric patterns. His technique was based on the Fourier series expansion. In 1943, Schelkunoff [2] utilized the relation between the roots of a polynomial on the complex plane and the nulls of the radiation pattern. Dolph [3] and Riblet [4] used the Chebyshev polynomials and offered a control on the side lobe level (SLL) of the pattern. Woodward and Lawson [5, 6], Van der Maas [7] and Taylor [8, 9] applied the method of matching the pattern of a continuous source to a certain number of directions or to the ideal Chebyshev factor. Cheng and Ma [10] used the sampling of a line source and the Z-transforms.

The above-mentioned sample of studies and most of the ones described in the previous chapters assume that the spacing of array elements is uniform. In 1956, Unz [11–13] proposed a synthesis of non-uniformly spaced arrays. He formulated the relation between the element excitations and the complex Fourier expansion of the radiation pattern in a matrix form. Non-uniformly spaced arrays offer more degrees of freedom than uniform arrays of the same number of elements, and, obviously, they are expected to have better performance. King, Packard and Thomas [14] and DuHamel and Chadwick [15] presented non-uniform arrays for broadband applications. Harrington [16] used a perturbation technique to reduce the size of the side lobes of linear arrays. For arrays with non-uniform spacing larger than 1λ and for broadband conditions, Unz [17] and Bruce [18] formulated the appropriate theories. Also, arrays with uniform excitation and non-uniform spacing have been designed by Pokrovskii [19] and Brown [20]. In 1962, Unz [21] and Lo [22] applied a formulation with which they transformed a continuous distribution into a non-uniformly spaced array. It should be noticed that Unz [23] and Uzkov [24] were the first researchers to perform significant work on the synthesis of non-uniformly spaced arrays to apply the orthogonal method. Since 1974, Sahalos and his colleagues [25–50] have extended and generalized the orthogonal methods for array synthesis.

Orthogonal Methods for Array Synthesis: Theory and the ORAMA Computer Tool   John N. Sahalos
© 2006 John Wiley & Sons, Ltd

In this chapter, several approaches to the orthogonal methods and their applications will be presented.

4.2 Synthesis of Non-uniformly Spaced Linear Arrays: The Matrix Inversion Method

The progressive improvement of feed networks and the reduction of their cost have permitted the use of complex and efficient antenna arrays. Antenna arrays contain, in general, arbitrarily oriented and non-uniformly spaced elements. Figure 4.1 presents a simple form of an antenna array with phase and amplitude control.

Let us first consider a non-uniformly spaced linear array with N elements (see Fig. 4.2). If the element positions are pre-assigned and the desired radiation pattern is given, the problem of synthesis is with the determination of the excitations I_n of the elements.

The array factor is of the following form:

$$AF(\theta) = \sum_{n=1}^{N} I_n e^{j\beta d_n \cos\theta} \tag{4-1}$$

We transform the independent variable θ into Equation (4-1) by making the following substitutions:

$$\left.\begin{aligned} u &= \pi \cos\theta \\ x_n &= \frac{d_n}{\lambda/2} \end{aligned}\right\} \tag{4-2}$$

Equation (4-1) is transformed into

$$AF(u) = \sum_{n=1}^{N} I_n e^{j\beta x_n u} = \sum_{n=1}^{N} I_n \phi_n(u) \tag{4-3}$$

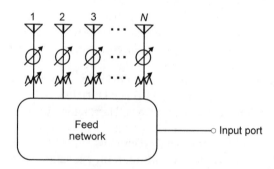

Figure 4.1   An antenna array with phase and amplitude control

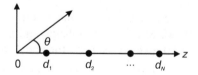

Figure 4.2   The geometry of a non-uniform linear array

The difficulty in the determination of the excitations comes from the fact that the array factor is expanded in a non-orthogonal set of functions. This is a problem that has been analysed by Kantorovich and Krylov [51].

One method of evaluating the excitations is to take a complete set of functions $\{\psi_n(u)\}$ and inner multiply both sides of Equation (4-3) by $\psi_n(u)$. One will then obtain a system of N linear equations of the following form:

$$\sum_{n=1}^{N} I_n < \phi_n(u), \psi_1(u) > = < AF(u), \psi_1(u) >$$

$$\sum_{n=1}^{N} I_n < \phi_n(u), \psi_2(u) > = < AF(u), \psi_2(u) > \qquad (4\text{-}4)$$

$$\cdots\cdots\cdots\cdots\cdots\cdots\cdots\cdots\cdots\cdots\cdots\cdots$$

$$\sum_{n=1}^{N} I_n < \phi_n(u), \psi_N(u) > = < AF(u), \psi_N(u) >$$

It is noticed that a suitable choice of the set of functions $\{\psi_n(u)\}$ is important.

By solving the Equation (4-4), we can derive the excitations I_n. The theory given above has been first developed by Unz [11–13] and involves matrix inversion. If the matrix of the system Equation (4-4) is singular, then the inverse does not exist. Also, if the matrix is 'almost singular', a solution can be obtained under certain conditions. The choice of $\{\psi_n(u)\}$ is critical for this aim. In [52], there is an interesting discussion on the stability of the matrix inversion problems in antennas.

4.3 Synthesis of Non-uniformly Spaced Linear Arrays: The Orthogonal Method

The determination of the excitations I_i of a non-uniformly spaced linear array can be achieved by using the orthogonal method. Suppose that the desired array factor $AF(u)$ is given. From Equation (4-3), it is obvious that the array factor is composed of basis functions $\phi_i(u) = e^{jx_iu}$ that, in general, are not orthogonal. With a method analogous to that of Gram-Schmidt [53–55], we can derive an orthogonal basis $\{\psi_i(u)\}$. The procedure has been discussed in detail by Unz [23], Uzkov [24] and Kantorovich and Krylov [51]. The orthogonal set of functions is of the following form:

$$\psi_1(u) = \frac{\phi_1(u)}{\langle \phi_1(u), \phi_1(u) \rangle^{\frac{1}{2}}} \qquad (4\text{-}5)$$

$$\psi_n(u) = \frac{\phi_n(u) - \sum_{j=1}^{n-1} \langle \phi_n(u), \psi_j(u) \rangle \psi_j(u)}{\left\{ \int_{-\pi}^{\pi} \left[\phi_n(u) - \sum_{j=1}^{n-1} \langle \phi_n(u), \psi_j(u) \rangle \psi_j(u) \right]^2 du \right\}^{\frac{1}{2}}} \qquad (4\text{-}6)$$

$\psi_n(u)$ is the orthonormalized function expressed by the equation

$$\psi_n(u) = \sum_{i=1}^{n} C_i^{(n)} \phi_i(u) \tag{4-7}$$

The symbol $\langle \phi_n(u), \psi_j(u) \rangle$ represents the inner product, which is given by

$$\langle \phi_n(u), \psi_j(u) \rangle = \int_{-\pi}^{\pi} \phi_n(u) \psi_j^*(u) \, du \tag{4-8}$$

From Equations (4-5) to (4-7), the coefficients $C_j^{(n)}$ can be found to be [25]

$$\left. \begin{aligned} C_n^{(n)} &= \frac{1}{D_n} \\ C_k^{(n)} &= -\frac{2\pi}{D_n} \sum_{j=k}^{n-1} \sum_{i=1}^{j} C_k^{(j)} C_i^{(j)} k_{in} \\ D_n &= \left\{ 2\pi - 4\pi^2 \sum_{j=1}^{n-1} \left(\sum_{i=1}^{j} C_i^{(j)} k_{in} \right)^2 \right\}^{1/2} \\ k_{in} &= \frac{\sin \pi (x_i - x_n)}{\pi (x_i - x_n)} \end{aligned} \right\} \tag{4-9}$$

Equation (4-3) is modified to

$$AF(u) = \sum_{i=1}^{N} B_i \psi_i(u) \tag{4-10}$$

where

$$B_i = \langle AF(u), \psi_i(u) \rangle \tag{4-11}$$

From Equations (4-3), (4-7) and (4-10), the excitations I_i result as follows [25]:

$$I_i = \sum_{j=i}^{N} B_j C_i^{(j)} \tag{4-12}$$

The above procedure proves that for a given non-uniformly spaced linear array, the element excitations for a desired array factor can be achieved. The accuracy of the approximation depends on the number and the position of the array elements. The whole synthesis follows five steps:

1. Definition of the position of the elements
2. Calculation of the coefficients $C_j^{(i)}$
3. Evaluation of the desired pattern $AF(u)$
4. Calculation of the quantities B_i.
5. Calculation of the excitations I_i of the array elements.

In the literature, there are several interesting studies about the well- and ill-conditioned problems in conjunction with the Gram-Schmidt procedure.

4.3.1 Examples of Synthesis of Linear Arrays with the Orthogonal Method

4.3.1.1 Optimum Directivity

This paragraph is devoted to examples of the synthesis of linear arrays with optimum directivity. Linear arrays are of great importance to antenna designers. For an array to present maximum directivity, $AF(u)$ must be an impulse function in the direction in which the maximum of the emission is sought.

$AF(u)$ can thus be approximated by Dirac's delta function, that is,

$$AF(u) = \delta(u - u_0) \tag{4-13}$$

Equations (4-11) and (4-13) combine to give

$$B_i = \psi_i^*(u_0) \tag{4-14}$$

Thus, from Equations (4-10) and (4-14), we have

$$AF(u_0) = \sum_{i=1}^{N} B_i B_i^* \tag{4-15}$$

Maximum directivity is

$$D(u_0) = \frac{|AF(u_0)|^2}{\frac{1}{2\pi} \int_{-\pi}^{\pi} |AF(u)|^2 \, du} = 2\pi \frac{\left| \sum_{i=1}^{N} B_i B_i^* \right|^2}{\int_{-\pi}^{\pi} \sum_{i=1}^{N} \sum_{j=1}^{N} B_i B_j^* \psi_i(u) \psi_j^*(u) \, du} \tag{4-16}$$

Finally, from Equation (4-16), we have

$$D(u_0) = 2\pi \sum_{i=1}^{N} B_i B_i^* = 2\pi \sum_{i=1}^{N} \psi_i(u_0) \psi_i^*(u_0) \tag{4-17}$$

The numerical computation of maximum directivity has been performed for the same arrays discussed by Tai [56] and Lo et al. [57]. Results showing maximum broadside directivity versus the element spacing of a uniformly spaced array are presented in Fig. 4.3 for $N = 3, 7, 11, 15$ and 19 elements. The results are in close agreement with the graphs of [56].

Also, results showing maximum end-fire directivity versus the element spacing of the above array are given. It can be seen in Fig. 4.4 that directivity approaches N^2 as $d/\lambda \to 0$. The results follow the same behaviour as the one given in [57].

Figures 4.5 and 4.6 present maximum directivity versus angle of maximization for $d/\lambda = 0.4$ and 0.2 respectively.

It is noticed that the greatest directivity occurs at end-fire direction. Also, for most angles, directivity improves only by a small value as the size of the array increases. Finally, the calculated scan patterns for a linear array with $N = 11$ elements (see Fig. 4.7) and the 3-D plot for a pattern with maximum at $\theta_{\max} = 55°$ (see Fig. 4.8) are given. In both cases, $d/\lambda = 0.2$.

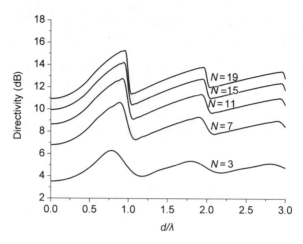

Figure 4.3   Maximum directivity of uniformly spaced broadside linear arrays for $N = 3, 7, 11, 15$ and 19 elements versus inter-element distance

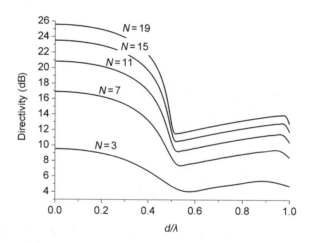

Figure 4.4   Maximum directivity of uniformly spaced end-fire linear arrays for $N = 3, 7, 11, 15$ and 19 elements versus inter-element distance

4.3.1.2 Chebyshev Patterns

The Chebyshev polynomials are attractive to antenna array designers because of their equal ripples. In Chapter 2, an analysis was given of the design possibilities of Chebyshev patterns. Our aim here is to give some useful numerical procedures that are helpful in the orthogonal method. We start with the array factor, which is

$$AF(u) = AF(x)$$

$$x = \frac{u}{\pi}, -1 \leq x \leq 1$$

(4-18)

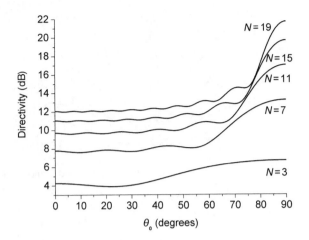

Figure 4.5 Maximum directivity versus angle of maximization for linear arrays ($N = 3, 7, 11, 15$ and 19) with $d/\lambda = 0.4$. It is $\theta_0 = 90° - \theta$

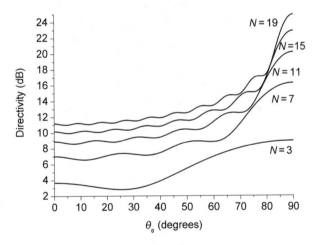

Figure 4.6 Maximum directivity versus angle of maximization for linear arrays ($N = 3, 7, 11, 15$ and 19) with $d/\lambda = 0.2$. It is $\theta_0 = 90° - \theta$

To find the excitations we start from B_n:

$$B_n = \langle AF(u), \psi_n(u) \rangle = \langle AF(x), \psi_n(x) \rangle \tag{4-19}$$

Equation (4-19) is written as

$$\langle AF(x), \psi_n(x) \rangle = \pi \int_{-1}^{1} AF(x) \sum_{i=1}^{n} C_i^{(n)} e^{-j(x_i \pi)x} \, dx \tag{4-20}$$

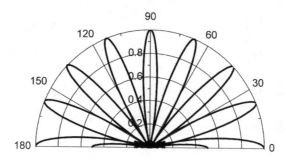

Figure 4.7   Scan patterns of a linear array with $N = 11, d/\lambda = 0.2$ and optimum directivity

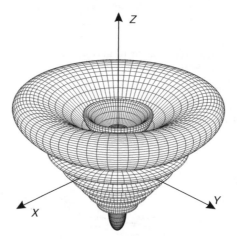

Figure 4.8   3-D plot of a pattern of a linear array with $N = 11, d/\lambda = 0.2$ and optimum directivity at $\theta_{max} = 55°$

We define the above sum of complex functions of x as

$$G_n(x) = AF(x) \sum_{i=1}^{n} C_i^{(n)} e^{-j(x_i\pi)x} \tag{4-21}$$

$G_n(x)$ can be approximated by using the sum [29]

$$G_n(x) \simeq \sum_{r=0}^{M} {}'' Q_r^{(n)} T_r(x) \tag{4-22}$$

where $T_r(x)$ is the Chebyshev polynomial of rth order and

$$Q_r^{(n)} = \frac{2}{M} \sum_{s=0}^{M} {}'' G_n\left(\cos\frac{\pi s}{m}\right) \cos\frac{\pi r s}{m} \tag{4-23}$$

It is known that [29]

$$\int_{-1}^{1} T_{2k}(x)\, dx = -\frac{2}{4k^2 - 1}$$

$$\int_{-1}^{1} T_{2k+1}(x)\, dx = 0$$

(4-24)

From Equations (4-20) to (4-24), it can be found that

$$\langle AF(x), \psi_n(x) \rangle = \pi \int_{-1}^{1} G_n(x)\, dx = B_n$$

(4-25)

or

$$B_n = -\frac{4\pi}{M} \sum_{k=0}^{M/2}{}'' \sum_{s=0}^{M}{}'' G_n \left(\cos \frac{\pi s}{M} \right) \cos \frac{2\pi k s}{M} / (4k^2 - 1)$$

(4-26)

We substitute Equation (4-21) in Equation (4-26) and get

$$B_n = -\frac{4\pi}{M} \sum_{k=0}^{M/2}{}'' \sum_{s=0}^{M}{}'' \sum_{i=1}^{n} C_i^{(n)} AF \left(\cos \frac{\pi s}{M} \right) e^{-j(x_i \pi) \cos \frac{\pi s}{M}} \cos \frac{2\pi k s}{M} / (4k^2 - 1)$$

(4-27)

(The symbol $\sum_{r=0}^{M}{}''$ represents the sum where the first and last coefficients are multiplied by $^1/_2$.

If M is odd, then $\dfrac{M}{2}$ is the $\dfrac{M-1}{2}$ and $\sum_{r=0}^{M/2}{}''$ has only the first coefficient multiplied by $1/2$)

From Equations (4-27) and (4-12), the excitations of the elements are derived.

Now, if the desired array factor $AF(x)$ is a Chebyshev polynomial of the form $T_L(x)$, then Equation (4-22) is an equation without any approximation. It is noticed that Equations (4-22) and (4-27) can be extended for 2-D and 3-D arrays.

I. Chebyshev Arrays with a General Independent Variable

Let us now look at the independent variable in the general form given in Equation (4-25), which is given as

$$x = a \cos(\beta d \cos \theta + \alpha) + b$$

(4-28)

To derive the coefficients a, b, α we give x three desired values. One of them must be x_0, ($|x_0| > 1$), which defines the SLL of the pattern. The other two are mainly chosen between -1 and $+1$ for two different angles of θ. Next, some interesting results are presented.

1. End-fire Chebyshev Arrays (SLL = −20 dB, N = 11 and d/λ = 0.25)

The Chebyshev polynomial is $T_5(x)$, and the values of the variable are

$$x_1 = x_0 = 1.1846, \theta_1 = 0°$$

$$x_2 = -1.000, \theta_2 = 180°$$

(4-29)

$$x_3 = 0.0, \theta_3 \cong 93.12°$$

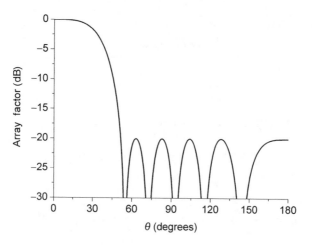

Figure 4.9 An end-fire Chebyshev pattern of an array with $N = 11, d/\lambda = 0.25$ and SLL $= -20$ dB for $\theta_3 = 93.12°$

The pattern for $\theta_3 = 93.12°$ is given in Fig. 4.9.
We change the values of the variables as follows:

$$x_1 = x_0 = 1.1846, \theta_1 = 0°$$
$$x_2 = 1.000, \theta_2 = 180°$$
$$x_3 = 0.0, 51.53° \leq \theta_3 \leq 137.72°$$
(4-30)

The pattern will also change and, for $\theta_3 = 51.53°$, it is given in Fig. 4.10.
It is observed that the pattern in Fig. 4.10 is more directive than that in Fig. 4.9. If one chooses a value between the bounds given in Equation (4-30) as θ_3, then the pattern

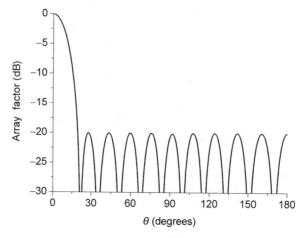

Figure 4.10 An end-fire Chebyshev pattern of an array with $N = 11, d/\lambda = 0.25$ and SLL $= -20$ dB for $\theta_3 = 51.53°$

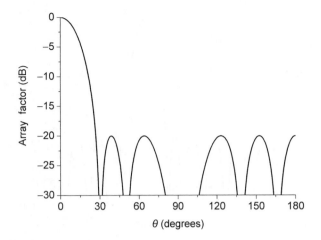

Figure 4.11 An end-fire Chebyshev pattern of an array with $N = 11, d/\lambda = 0.25$ and SLL $= -20$ dB for $\theta_3 = 90°$

becomes less directive than that with a value in the bounds. An example is given in Fig. 4.11 for $\theta_3 = 90°$.

We change the values of the variables again as follows:

$$x_1 = x_0 = 1.1846, \theta_1 = 0°$$
$$x_2 = 1.000, \theta_2 = 180° \qquad (4\text{-}31)$$
$$x_3 = 1.000, 19.3° \leq \theta_3 \leq 180°$$

The pattern will change and, for different values of θ_3, we will have different patterns (see Fig. 4.12).

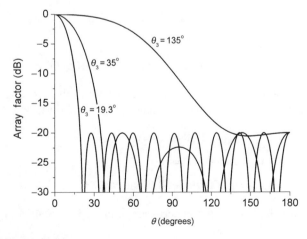

Figure 4.12 End-fire Chebyshev pattern of an array with $N = 11, d/\lambda = 0.25$ and SLL $= -20$ dB for three different values of θ_3

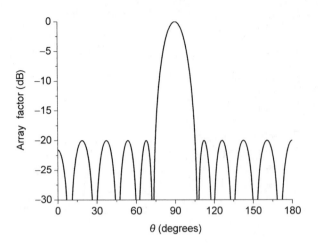

Figure 4.13 A broadside Chebyshev pattern of an array with $N = 11, d/\lambda = 0.25$ and SLL $= -20$ dB for $\theta_3 = 45.66°$

2. Broadside Chebyshev Arrays (SLL $= -20$ dB, $N = 11$ and $d/\lambda = 0.25$)
For broadside arrays, the Chebyshev polynomial is again $T_5(x)$, and the values of the variable are

$$x_1 = x_0 = 1.1846, \theta_1 = 90°$$
$$x_2 = -1.000, \theta_2 = 180° \tag{4-32}$$
$$x_3 = 0.0, \theta_3 \cong 45.66° \text{ and } \theta_3 \cong 134.19°$$

For the permitted values of θ_3, the patterns are similar. In Fig. 4.13, a pattern is given for $\theta_3 = 45.66°$.
We change the values of the variables as follows:

$$x_1 = x_0 = 1.1846, \theta_1 = 90°$$
$$x_2 = -1.000, \theta_2 = 180° \tag{4-33}$$
$$x_3 = 1.000, \theta_3 \cong 74.66° \text{ and } \theta_3 \cong 105.25°$$

For the permitted values of θ_3, the patterns are similar to those of Fig. 4.13.
Finally, we change the values of the variables again as follows:

$$x_1 = x_0 = 1.1846, \theta_1 = 90°$$
$$x_2 = 0.000, \theta_2 = 180° \tag{4-34}$$
$$x_3 = 1.000, \theta_3 \cong 68.9° \text{ and } \theta_3 \cong 111.11°$$

For the permitted values of θ_3, the patterns are similar. In Fig. 4.14, a pattern is given for $\theta_3 = 68.9°$.
The above given values of x_1, x_2 and x_3 are not the only ones. One could select a lot of suitable sets that follow certain constraints.

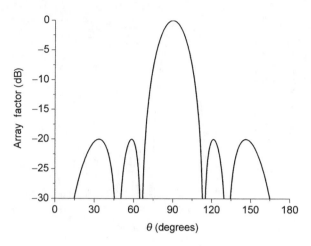

Figure 4.14 A broadside Chebyshev pattern of an array with $N = 11, d/\lambda = 0.25$ and SLL $= -20$ dB for $\theta_3 = 68.9°$

3. Intermediate Chebyshev Arrays (SLL $= -20$ dB, $N = 11$ and $d/\lambda = 0.25$)
For intermediate arrays with $N = 11$, the Chebyshev polynomial is again $T_5(x)$. If the maximum of the pattern is desired to be at $60°$, then the values of the variables are

$$x_1 = x_0 = 1.1846, \theta_1 = 60°$$
$$x_2 = -1.000, \theta_2 = 180° \tag{4-35}$$
$$x_3 = 0.0, \theta_3 \cong 116.93°$$

In Fig. 4.15, a pattern is given for $\theta_3 = 116.93°$.

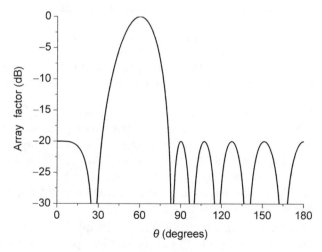

Figure 4.15 A Chebyshev pattern of an intermediate array with $N = 11, d/\lambda = 0.25$ and SLL $= -20$ dB for $\theta_3 = -116.93°$

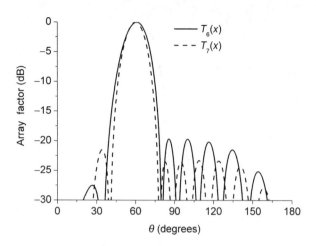

Figure 4.16 Chebyshev patterns of an intermediate array of 11 collinear small dipoles with $T_6(x)$ and $T_7(x)$ for $N = 11$, $d/\lambda = 0.25$ and SLL ≤ -20 dB

Table 4.1 Excitation currents of intermediate arrays with Chebyshev patterns $T_5(x)$, $T_6(x)$ and $T_7(x)$ for $N = 11$, $d/\lambda = 0.25$ and SLL ≤ -20 dB

Element number	Excitation coefficients		
	$T_5(x)$	$T_6(x)$	$T_7(x)$
1	$-0.2235 - j0.2220$	$0.0314 - j0.1448$	$-0.0240 + j0.0057$
2	$0.2350 - j0.0006$	$-0.1052 + j0.4410$	$0.1212 - j0.0151$
3	$-0.4827 + j0.4846$	$0.0939 - j0.8088$	$-0.3269 + j0.0256$
4	$-0.6864 - j0.5201$	$-0.0344 + j0.9994$	$0.6205 - j0.0311$
5	$0.7076 + j0.7066$	$-0.0634 - j0.6749$	$-0.8910 + j0.0201$
6	$-0.6546 + j0.0000$	$0.1466 + j0.0000$	$1.0000 - j0.0000$
7	$0.7076 - j0.7066$	$-0.0634 + j0.6749$	$-0.8910 - j0.0201$
8	$-0.6864 + j0.5201$	$-0.0344 - j0.9994$	$0.6205 + j0.0311$
9	$-0.4827 - j0.4846$	$0.0939 + j0.8088$	$-0.3269 - j0.0256$
10	$0.2350 + j0.0006$	$-0.1052 - j0.4410$	$0.1212 + j0.0151$
11	$-0.2235 + j0.2220$	$0.0314 + j0.1448$	$-0.0240 - j0.0057$

Different results are obtained if we choose a different set of x_1, x_2 and x_3.

It is interesting to see how the orthogonal method works if, instead of $T_5(x)$, we use a $T_m(x)$, $m > 5$ for $N = 11$. In Fig. 4.16, the patterns of a linear array of collinear small dipoles with $N = 11$ for $T_6(x)$ and $T_7(x)$ are presented. Also, in Table 4.1, the excitations of the elements of the array for $T_m(x)$, $m = 5, 6, 7$ are given.

It is obvious that the orthogonal method can give solutions to Chebyshev arrays with different orders and the same number of elements.

4. Chebyshev Arrays for a Desired HPBW (SLL = -20 dB, $N = 11$)
In a Chebyshev array, if, besides the SLL, a certain half-power bandwidth (HPBW) is desirable, one can solve the problem by using the appropriate angles. We start with a broadside array, where a certain value of HPBW = $12°$ is desirable. We choose x_1, x_2

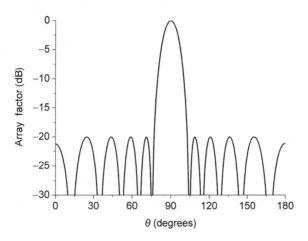

Figure 4.17 Radiation pattern of a broadside Chebyshev array for $N = 11, d/\lambda = 0.4$ and HPBW $= 12°$

and x_3 for a $T_5(x)$ pattern as follows:

$$x_1 = x_0 = 1.1846, \theta_1 = 90°$$

$$x_2 = 1.1432, \theta_2 = 90° + \frac{\text{HPBW}}{2} \tag{4-36}$$

$$x_3 = 0.000, \theta_3 = 51.8°$$

It is noticed that x_2 is the value where the level is 3 dB less than the main lobe value of its maximum. Figure 4.17 presents the pattern of a broadside array for $d/\lambda = 0.4$ and HPBW $= 12°$.

We move on to an end-fire array, where, again, a certain value of HPBW $= 30°$ is desirable. We choose x_1, x_2 and x_3 as follows:

$$x_1 = x_0 = 1.1846, \theta_1 = 0°$$

$$x_2 = 1.1432, \theta_2 = \frac{\text{HPBW}}{2} \tag{4-37}$$

$$x_3 = 0.000, \theta_3 = 61°$$

It is noticed again that x_2 is the value where the main lobe level is 3 dB less than the main lobe value of its maximum. Figure 4.18 presents the pattern of an end-fire array for $d/\lambda = 0.4$ and HPBW $= 30°$.

Finally, we move on to an intermediate array. For a certain value of HPBW $= 20°$, we choose x_1, x_2 and x_3 as follows:

$$x_1 = x_0 = 1.1846, \theta_1 = 60°$$

$$x_2 = 1.1432, \theta_2 = 60° + \frac{\text{HPBW}}{2} \tag{4-38}$$

$$x_3 = 1.000, \theta_3 = 80.75°$$

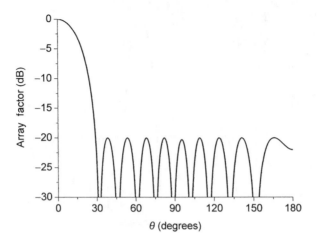

Figure 4.18 Radiation pattern of an end-fire Chebyshev array for $N = 11, d/\lambda = 0.4$ and HPBW $= 30°$

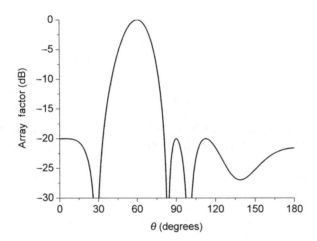

Figure 4.19 Radiation pattern of an intermediate Chebyshev array for $N = 11, d/\lambda = 0.4$ and HPBW $= 20°$

It is noticed again that x_2 is the value where the level is 3 dB less than the main lobe value of its maximum. Figure 4.19 presents the pattern of an intermediate array for $d/\lambda = 0.4$ and HPBW $= 20°$.

The orthogonal method can be applied for patterns other than $T_5(x)$. If a $T_7(x)$ pattern is chosen, the values x_1, x_2 and x_3 will be

$$x_1 = x_0 = 1.0928, \theta_1 = 60°$$
$$x_2 = 1.072, \theta_2 = 60° + \frac{\text{HPBW}}{2} \tag{4-39}$$
$$x_3 = 1.000, \theta_3 = 38.5°$$

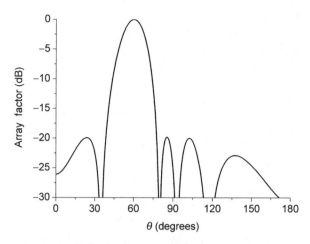

Figure 4.20 Radiation pattern of an intermediate Chebyshev array for $N = 11, d/\lambda = 0.4$, maximum at $60°$ and HPBW $= 17°$. Chebyshev polynomial is $T_7(x)$ instead of $T_5(x)$

Figure 4.20 presents the pattern of an intermediate array for $d/\lambda = 0.4$ and HPBW $= 17°$.

Finally, it is possible to use a pattern of up to $T_{10}(x)$ with the following values of x_1, x_2 and x_3:

$$x_1 = x_0 = 1.0451, \theta_1 = 60°$$

$$x_2 = 1.0352, \theta_2 = 60° + \frac{\text{HPBW}}{2} \qquad (4\text{-}40)$$

$$x_3 = 1.000, \theta_3 = 38.8°$$

Figure 4.21 presents the pattern of an intermediate array for $d/\lambda = 0.4$ and HPBW $= 17°$.

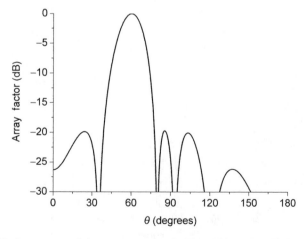

Figure 4.21 Radiation pattern of the same array as in Fig. 4.20, with a Chebyshev pattern $T_{10}(x)$ instead of $T_5(x)$

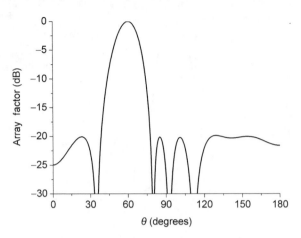

Figure 4.22   Radiation pattern of an intermediate Chebyshev array for $N = 10$, $d/\lambda = 0.46$, maximum at $60°$ and HPBW $= 17°$. The $T_5(x)$ polynomial is used

For an array with an even number of elements, one can use the orthogonal method for any of the above cases. Let us suppose that we have $N = 10$ elements and it is desirable to have a $T_5(x)$ pattern. Normally, we need $N = 11$ elements. For a smaller number, we increase the inter-element distance. So, for values of x_1, x_2 and x_3 given in Equation (4-38) and $d/\lambda = 0.46$, we have the desired pattern for HPBW $= 17°$, $\theta_1 = 60°$ (see Fig. 4.22).

II. Special Cases of End-fire Chebyshev Arrays

Special cases of end-fire arrays can be given for certain values of x_1, x_2 and x_3. In Table 2.4, the four most common cases with the permitted values of d/λ and the corresponding ones of a, b and α have been presented.

Figures (4.23) to (4.26) present four examples of patterns for the corresponding cases of Table 2.4.

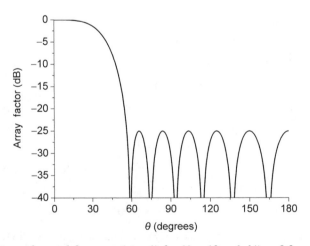

Figure 4.23   Pattern of an end-fire array (case 1) for $N = 13$ and $d/\lambda = 0.2$ and SLL $= -25$ dB

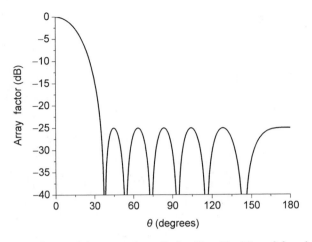

Figure 4.24 Pattern of an end-fire array (case 2) for $N = 13$, $d/\lambda = 0.2$ and SLL $= -25$ dB

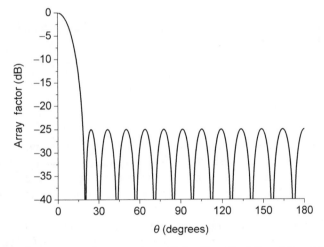

Figure 4.25 Pattern of an end-fire array (case 3) for $N = 13$, $d/\lambda = 0.2$ and SLL $= -25$ dB

III. Dolph-Chebyshev Arrays

In the Dolph formulation, it is usually desirable to have broadside arrays. However, the independent variable is of the following form:

$$x = x_0 \cos \frac{\psi}{2}$$
$$\psi = \beta d \cos \theta + \alpha \tag{4-41}$$

If the desired maximum is at an angle θ_0, the phase difference between the elements is

$$\alpha = -\beta d \cos \theta_0 \tag{4-42}$$

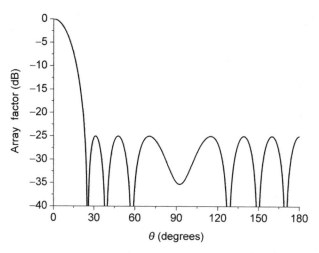

Figure 4.26   Pattern of an end-fire array (case 4) for $N = 13$, $d/\lambda = 0.2$ and SLL $= -25$ dB, and HPBW $= 20°$

It is noticed that there exists an upper (d_{max}) and a lower (d_{min}) bound for the permitted values of d/λ. For $\theta_0 = 90°$, these values are

$$\frac{d_{max}}{\lambda} = 1 - \frac{\cos^{-1}(1/x_0)}{\pi}$$

$$\frac{d_{min}}{\lambda} = \frac{\cos^{-1}(x_{min}/x_0)}{\pi}$$

(4-43)

where x_{min} is the position of the first null of the Chebyshev polynomial.

For $N = 11$ elements, the desired pattern is $T_{10}(x)$ and the permitted values for SLL $= -20$ dB are $0.136 \le d/\lambda \le 0.906$. It is noticed that for each distance d/λ, a different HPBW is achieved. Figure 4.27 presents the pattern of the above array for $d/\lambda = 0.5$.

If it is desirable to have a $T_{11}(x)$ pattern with the same elements as above and the same SLL, it is possible to apply the orthogonal method. In fact, the $T_{11}(x)$ pattern needs 12 elements. The distance d/λ for 11 elements is chosen in such a way so as to give the same array size as that of 12 elements. So, an array with $N = 12$ and $d/\lambda = 0.5$ is of the same length (5.5λ) as an array with $N = 11$ and $d/\lambda = 0.55$. Figure 4.28 presents the pattern $T_{11}(x)$ for 11 and 12 elements. It is observed that the patterns are approximately the same.

In Table 4.2, the excitation amplitudes are given for the above two arrays.

Table 4.2   The excitation amplitudes of Dolph-Chebyshev $T_{11}(x)$ pattern for $N = 11$ and 12 elements

	n	1,12	2,11	3,10	4,9	5,8	6,7
$N = 12$	A_n	0.713	0.529	0.789	0.845	0.976	1.000
$N = 11$	n	1, 11	2, 10	3, 9	4, 8	5, 7	6
	A_n	0.667	0.569	0.733	0.875	0.969	1.000

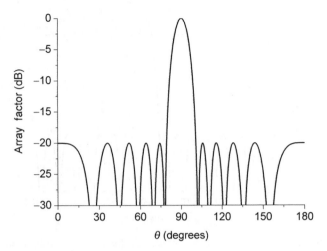

Figure 4.27 Pattern $T_{10}(x)$ of a classical Dolph-Chebyshev array for $N = 11, d/\lambda = 0.5$ and SLL $= -20$ dB

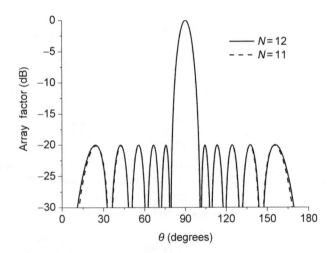

Figure 4.28 Pattern $T_{11}(x)$ of a classical Dolph-Chebyshev array for $N = 12, d/\lambda = 0.5$ and for $N = 11, d/\lambda = 0.55$ (SLL $= -20$ dB)

IV. Riblet Broadside Arrays

A Riblet array corresponds to $2n + 1$ elements for a pattern $T_n(x)$. The values x_1, x_2 and x_3 of the variable x are

$$x_1 = x_0 = a + b$$
$$x_2 = a\cos(\beta d - \alpha) + b \qquad (4\text{-}44)$$
$$x_3 = a\cos(\beta d \cos\theta_{\text{HP}} - \alpha) + b$$

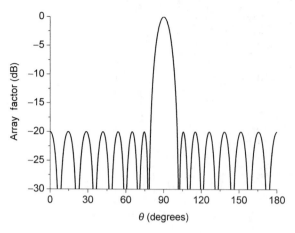

Figure 4.29　Pattern $T_7(x)$ of a classical Riblet array for an array with $N = 15, d/\lambda = 0.279$, HPBW $= 10°$ and SLL $= -20$ dB

The coefficients a, b, α are

$$a = \frac{x_0 - x_2}{1 - \cos(\beta d - \alpha)}$$

$$b = x_0 - a \tag{4-45}$$

$$\alpha = -\beta d \cos \theta_0$$

For $\theta_0 = 90°$ (broadside case), it is

$$\theta_{HP} = \sin^{-1} \left[\frac{1}{\beta d} \cos^{-1} \left(\frac{x_3 - b}{a} \right) \right] \tag{4-46}$$

The inter-element distance is $\le \lambda/2$ for $\theta_0 = 90°$. It is obvious that the HPBW of the array depends on the distance d/λ. From Equations (4-45) and (4-46), it can be found that

$$\cos(\beta d \cos \theta_{HP}) - k[1 - \cos(\beta d)] = 1 \tag{4-47}$$

where

$$k = \frac{x_3 - x_0}{x_0 - x_2} \tag{4-48}$$

Equation (4-47) offers the possibility of choice of either d/λ or HPBW. Figure 4.29 presents the pattern of a Riblet array for $N = 15$ elements. The desired HPBW is $10°$. From Equation (4-47), it is found that $d/\lambda = 0.279$.

4.3.1.3 Exponential Patterns

In this subsection, we will present four different exponential patterns. These are:

$$AF(u) = e^{-a^2 u^2} \cos(2bu), u = \pi \cos \theta$$

$$AF(u) = e^{-au}, u = \pi \cos \theta$$

$$AF(u) = e^{-au}, u = \pi \sin \theta \tag{4-49}$$

$$AF(\theta) = e^{-k\left(\theta - \theta_{in} - \frac{\theta_0}{2}\right)^2}$$

I. Pattern $AF(u) = e^{-a^2 u^2} \cos(2bu)$

This type of pattern has been first proposed by Unz [21] and is useful for the design of broadside arrays by offering two different choices. One is the choice of having a specific null of the main lobe and a desired SLL. The other is to have a specific null of the main lobe and a desired HPBW. If the null of the main lobe is at the angle θ_0, then it is found that

$$b = \frac{1}{4 \cos \theta_0} \tag{4-50}$$

If a certain SLL is desired, we solve the system of the following two equations:

$$\frac{d}{du} AF(u) = 0 \text{ at } u = u_{SLL} \tag{4-51}$$

and

$$e^{-a^2 u_{SLL}^2} \cos(2b u_{SLL}) = -R \tag{4-52}$$

where $R = 10^{SLL_{db}/20}$.

From Equations (4-51) and (4-52), we have the following two equations for a^2:

$$a^2 = -b \frac{\tan(2b u_{SLL})}{u_{SLL}}$$
$$a^2 = -\frac{1}{u_{SLL}^2} \ln \left(\frac{-R}{\cos(2b u_{SLL})} \right) \tag{4-53}$$

By combining Equation (4-53), we obtain an equation for u_{SLL}, which is

$$z e^{\frac{\sqrt{1-z^2} \cos^{-1}(z)}{2z}} + R = 0 \tag{4-54}$$

where

$$z = \cos(2b u_{SLL}), -1 \leq z < 0 \tag{4-55}$$

Equation (4-54) is solved numerically for z and, from Equation (4-55), u_{SLL} is found. Next, a^2 is derived from (4-53). Finally, HPBW is calculated by solving numerically the equation $AF(u) = \frac{\sqrt{2}}{2}$ in the range $\left(0, \frac{\pi}{4b}\right)$.

If a certain value of HPBW is desired, a^2 is derived by using the equation

$$AF(u_{HP}) = \frac{\sqrt{2}}{2} \tag{4-56}$$

where $u_{HP} = \pi \sin \left(\frac{HPBW}{2} \right)$.

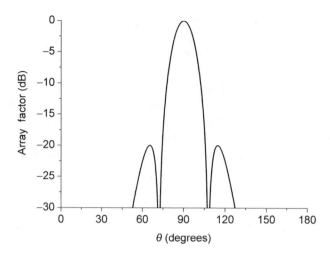

Figure 4.30   Pattern of a linear array with main lobe null at $\theta = 72°$ and SLL $= -20$ dB, where $N = 12$ and $d/\lambda = 0.45$

The value of a^2 is

$$a^2 = \frac{\ln[\sqrt{2}\cos(2bu_{HP})]}{u_{HP}^2} \qquad (4\text{-}57)$$

Let us now see two examples of an array with $N = 12$ elements and $d/\lambda = 0.45$. In the first example, it is desirable to have the main lobe null at $\theta = 72°$ and SLL $= -20$ dB. The coefficients of the pattern and the HPBW are

$$a = 0.981$$
$$b = 0.809 \qquad (4\text{-}58)$$
$$\text{HPBW} = 14°$$

Figure 4.30 presents the pattern of the array.

In the second example, the main lobe null is chosen to be at $\theta = 70°$ and HPBW $= 15°$. The coefficients of the pattern and the SLL are

$$a = 0.960$$
$$b = 0.731 \qquad (4\text{-}59)$$
$$\text{SLL} = -22.45 \text{ dB}$$

By using the orthogonal method with a non-uniformly spaced array, the same problem can be solved. Suppose that the array has $N = 9$ elements at symmetric positions around the central one. The positions are given in Table 4.3.

Figure 4.31 presents the pattern of the array. As it is observed, the pattern follows the desired constraints.

Table 4.3 Positions in wavelengths of the elements of a linear array with $N = 9$ elements

n	1,9	2,8	3,7	4,6	5
d_n/λ	±2.012	±1.447	±0.953	±0.476	0.000

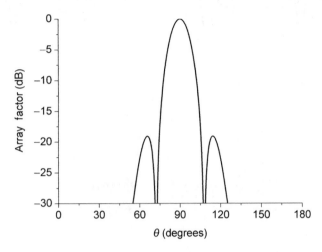

Figure 4.31 Pattern of a linear array with main lobe null at $\theta = 70°$ and HPBW $= 15°$. It is $N = 9$ and the position of the elements is given in Table 4.3

II. Pattern $AF(u) = e^{-au}, u = \pi \cos \theta$

This pattern comes from a broadside array with a parameter that determines the HPBW. The parameter a is found by

$$a = -\frac{\ln\left(\frac{\sqrt{2}}{2}\right)}{\pi \sin\left(\frac{\text{HPBW}}{2}\right)} \tag{4-60}$$

The desired pattern does not contain side lobes. An example is given in Figs. 4.32a–c where it is desirable to have a pattern with HPBW $= 10°$. Three different arrays with the same size of 8.4λ are used. These are:

$N = 15$ with $d/\lambda = 0.6$ (see Fig. 4.32a)
$N = 13$ with $d/\lambda = 0.7$ (see Fig. 4.32b)
$N = 12$ with $d/\lambda = 0.7636$ (see Fig. 4.32c)

It is observed that a smaller number of elements produce side lobes with a higher level.

III. Pattern $AF(u) = e^{-au}, u = \pi \sin \theta$

This pattern comes from an end-fire array with a parameter that determines HPBW. The parameter a is found by Equation (4-60). Figure 4.33 presents the pattern of an array with

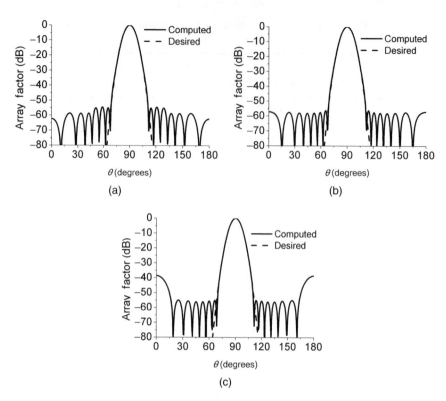

Figure 4.32 Pattern of a broadside array with HPBW $= 10°$. (a) $N = 15$ with $d/\lambda = 0.6$, (b) $N = 13$ with $d/\lambda = 0.7$ and (c) $N = 12$ with $d/\lambda = 0.7636$

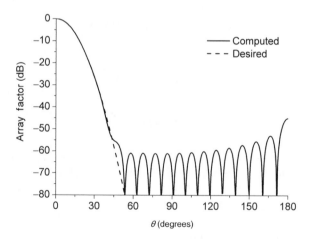

Figure 4.33 Pattern of an end-fire array with HPBW $= 20°$, $N = 18$ and $d/\lambda = 0.215$

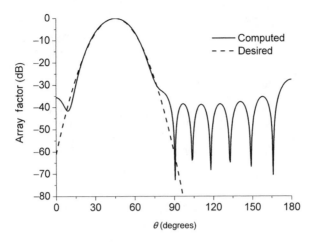

Figure 4.34 Pattern of a linear array with HPBW $= 20°$ at $\theta_{max} = 45°$, $N = 11$ and $d/\lambda = 0.28$

$N = 18$ elements and $d/\lambda = 0.215$ for HPBW $= 20°$. Here, it is more difficult to achieve a pattern without side lobes. Also, the rule of increasing the spacing d/λ as the number of elements decreases does not apply.

IV. Pattern $AF(\theta) = e^{-k\left(\theta - \theta_{in} - \frac{\theta_0}{2}\right)^2}$

This array factor is a one-lobe factor with maximum at a certain angle and a desired HPBW. It is

$$\theta_{max} = \theta_{in} + \frac{\theta_0}{2}$$

$$\text{HPBW} = \theta_0 \tag{4-61}$$

$$k = \frac{2\ln(2)}{\theta_0^2}$$

An example of an array with HPBW $= 20°$ at $\theta_{max} = 45°$, $N = 11$ and $d/\lambda = 0.28$ is given in Fig. 4.34.

4.3.1.4 Pulse Patterns

Pulse patterns represent the array factor as a set of pulses. The pulses can have the desired amplitudes in certain angular regions.

A typical example of an array with $N = 18$ and $d/\lambda = 0.215$ for the following pulse pattern is given in Fig. 4.35.

$$AF(\theta) = 1.0, \quad 0° \leq \theta \leq 15°$$

$$AF(\theta) = 0.1, \quad 25° \leq \theta \leq 45° \tag{4-62}$$

$$AF(\theta) = 0.1, \quad 160° \leq \theta \leq 180°$$

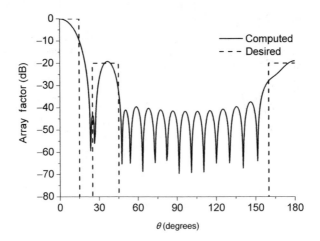

Figure 4.35 Pulse pattern of an array with $N = 18$ and $d/\lambda = 0.215$ and array factor given in Equation (4-62)

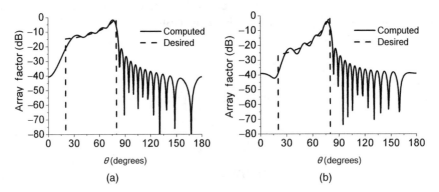

Figure 4.36 Patterns of a linear array with $N = 21$ elements and $d/\lambda = 0.5$ for $\theta_1 = 20°, \theta_2 = 80°$ and two different values $a = 1$ (left), $a = 1.8$ (right)

4.3.1.5 Inverse Trigonometric Patterns

Two different patterns of the following form will be presented here:

$$AF(\theta) = \begin{cases} \dfrac{1}{\cos^a(\theta)} & \theta_1 \leq \theta \leq \theta_2, \theta \neq 90°, 270° \\ 0 & \text{otherwise} \end{cases}$$

$$AF(\theta) = \begin{cases} \dfrac{1}{\sin^a(\theta)} & \theta_3 \leq \theta \leq \theta_4, \theta \neq 0°, 180° \\ 0 & \text{otherwise} \end{cases}$$

(4-63)

The exponential parameter of both array factors determines the slope of the main lobe. The patterns of an array with $N = 21$ elements and $d/\lambda = 0.5$ for $\theta_1 = 20°, \theta_2 = 80°$ and two different values $a = 1, a = 1.8$ are presented in Figs. 4.36a and b.

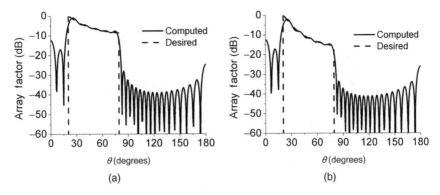

Figure 4.37 Patterns of a linear array with $N = 31$ elements and $d/\lambda = 0.375$ for $\theta_3 = 20°$, $\theta_4 = 80°$ and two different values $a = 1$ (left), $a = 1.8$ (right)

Also, the patterns of an array with $N = 31$ elements and $d/\lambda = 0.375$ for $\theta_3 = 20°$, $\theta_4 = 80°$ and two different values $a = 1$, $a = 1.8$ are presented in Figs. 4.37a and b.

4.3.1.6 Radar Patterns

Patterns useful for radar applications are presented as a combination of exponential, inverse trigonometric and pulse patterns. These are:

$$AF(\theta) = \begin{cases} \dfrac{1}{\cos^a(\theta)} & \theta_1 \le \theta \le \theta_2, \theta \neq 0°, 180° \\ e^{-k\left(\theta-\theta_{in}-\frac{\theta_0}{2}\right)^2} & \theta > \theta_2 \\ AF_1 & \theta_3 < \theta < \theta_4 \end{cases} \tag{4-64}$$

A typical example of a radar pattern of a linear array with $N = 31$ elements, $d/\lambda = 0.45$, $\theta_1 = 20°$, $\theta_2 = 80°$, $\theta_3 = 125°$, $\theta_4 = 180°$, $AF_1 = 0.018$, HPBW $= 15°$ and $a = 1.5$ is given in Fig. 4.38. It is obvious that the pattern follows the desired constraints.

4.3.1.7 Taylor/Bayliss Patterns by Iteration

The Taylor and Bayliss patterns, which have been taken by iteration (see Chapter 2), can be produced, by using the orthogonal method. Here, the equations of 2.5.5 are used in conjunction with the orthogonal method. A typical example, where two inner side lobes on each side of the main lobe are depressed, is given. A linear array is used with $N = 11$ elements; in equal space $d/\lambda = 0.45$. It is desirable to have the inner side lobes at -40 dB and the outer ones at -20 dB. Figure 4.39 presents the Taylor pattern.

The same pattern is observed using the orthogonal method for a non-uniformly spaced array. The positions of the elements are given in Table 4.4 (see Fig. 4.40).

Finally, in Fig. 4.41, a Bayliss pattern is given for $N = 13$ elements in $d/\lambda = 0.42$ and the same constraints in SLL as above.

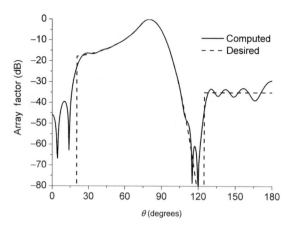

Figure 4.38 Radar pattern of a linear array with $N = 31$ elements, $d/\lambda = 0.45, \theta_1 = 20°$, $\theta_2 = 80°, \theta_3 = 125°, \theta_4 = 180°, AF_1 = 0.018$, HPBW $= 15°, a = 1.5$

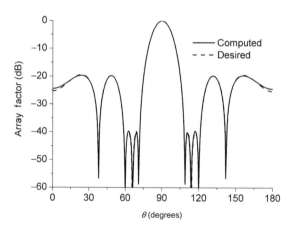

Figure 4.39 Taylor pattern of a linear array with $N = 11$ elements; in equal space $d/\lambda = 0.45$. The inner side lobes are at -40 dB and the outer ones are at -20 dB

Table 4.4 Positions in wavelengths of the elements of a linear array with $N = 11$

n	1,11	2,10	3,9	4,8	5,7	6
d_n/λ	±2.217	±1.727	±1.296	±0.870	±0.453	0.000

4.4 Synthesis of Non-uniformly Spaced Linear Arrays Subject to Constraints: The Constrained Orthogonal Method

The synthesis of a linear array under constraints is common in antenna engineering. Three different cases will be presented here. The first has to do with constraints on pattern nulls, the second with minimization of the pattern level in predefined angular regions and the

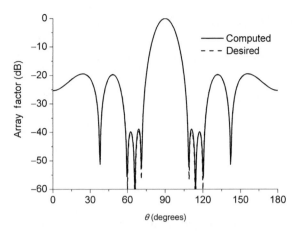

Figure 4.40 Taylor pattern of a linear array with $N = 11$ elements in the positions given in Table 4.4. The inner side lobes are at -40 dB and the outer ones are at -20 dB

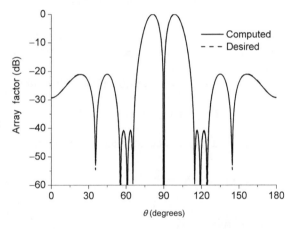

Figure 4.41 Bayliss pattern of a linear array with $N = 13$ elements, in $d/\lambda = 0.42$. The inner side lobes are at -40 dB and the outer ones are at -20 dB

third with constraints on certain SLLs. The orthogonal method is used in combination with the mean-square-error (MSE) criterion and the Lagrange technique.

It is well known that a useful criterion in array synthesis is the minimization of MSE. Let us assume a desired array factor $AF_d(u)$ that is achieved by a linear array (see Fig. 4.2) with the excitation currents I_n. The array factor is

$$AF(u) = \sum_{n=1}^{N} I_n \phi_n(u) \tag{4-65}$$

With the orthogonal method, Equation (4-65) is modified to

$$AF(u) = \sum_{n=1}^{N} B_n \psi_n(u) \tag{4-66}$$

The MSE between $AF_d(u)$ and $AF(u)$ is

$$\text{MSE} = \frac{1}{2\pi} \int_{-\pi}^{\pi} |AF_d(u) - AF(u)|^2 \, du = \frac{1}{2\pi} \int_{-\pi}^{\pi} \left|AF_d(u) - \sum_{n=1}^{N} B_n \psi_n(u)\right|^2 du \quad (4\text{-}67)$$

By using the orthogonality of $\psi_n(u)$, Equation (4-67) becomes

$$\text{MSE} = \frac{1}{2\pi} \left[\int_{-\pi}^{\pi} |AF_d(u)|^2 \, du - \sum_{n=1}^{N} (B_n^* \Gamma_n + B_n \Gamma_n^*) + \sum_{n=1}^{N} B_n^* B_n \right] \quad (4\text{-}68)$$

where

$$\Gamma_n = \int_{-\pi}^{\pi} AF_d(u) \psi_n^*(u) \, du \quad (4\text{-}69)$$

From Equation (4-67), it is obvious that MSE is a parabolic function with a minimum at the lower bound. To find the minimum we take

$$\frac{\partial \text{MSE}}{\partial B_n^*} = 0 \Rightarrow B_n - \Gamma_n \Rightarrow B_n = \langle AF_d(u), \psi_n(u) \rangle \quad (4\text{-}70)$$

Equation (4-70) proves that the minimization of MSE is equivalent to synthesis with the orthogonal method because it gives the same values for the excitations. We proceed now to synthesis under constraints.

It is desirable to minimize the MSE given in Equation (4-70), subject to the constraints that

$$f_m(\ldots, B_i^*, \ldots, B_j, \ldots) = 0, m = 1, \ldots, M \quad (4\text{-}71)$$

This constrained minimization can be accomplished by using the Lagrangian

$$L = \text{MSE} + \sum_{m=1}^{M} \lambda_m f_m(\ldots, B_i^*, \ldots, B_j, \ldots) \quad (4\text{-}72)$$

where λ_m is a Lagrange multiplier [57].

It is desirable to minimize L with respect to Equation (4-71).

$$\frac{\partial L}{\partial B_n^*} = 0 = \frac{\partial \text{MSE}}{\partial B_n^*} + \sum_{m=1}^{M} \lambda_m \frac{\partial f_m(\ldots, B_i^*, \ldots, B_j, \ldots)}{\partial B_n^*}, n = 1, \ldots, N \quad (4\text{-}73)$$

By solving Equations (4-71) and (4-73), we derive the values of λ_m, B_n.

4.4.1 Synthesis Under Null Constraints on Pattern Levels

If the constraints have to do with certain pattern nulls, then $f(u)_{u=u_m}$ are

$$f_m(\ldots, B_i^*, \ldots, B_j, \ldots) = \sum_{n=1}^{N} B_n^* \psi_n^*(u_m) = 0, m = 1, \ldots, M \quad (4\text{-}74)$$

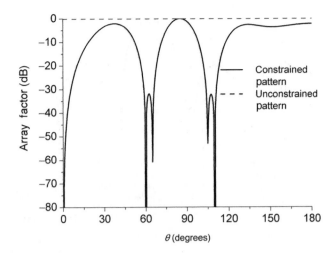

Figure 4.42 Pattern of a linear array of $N = 8$ elements and $d/\lambda = 0.4$ with nulls at $0°, 60°, 110°$

From Equations (4-73) and (4-74), we have

$$B_n = \Gamma_n + \sum_{m=1}^{M} \lambda_m \psi_n^*(u_m)$$

$$\sum_{j=1}^{M} \lambda_j \left(\sum_{n=1}^{N} \psi_n(u_j) \psi_n^*(u_m) \right) = \sum_{n=1}^{N} \Gamma_n^* \psi_n^*(u_m), m = 1, \ldots, M \tag{4-75}$$

By solving Equation (4-75), one can find λ_j, B_n and can finally solve the problem.

It is interesting to give some typical examples. We start with one where it is desirable to suppress the interference from the directions $0°$, $60°$ and $110°$ to a linear array with $N = 8$ elements and $d/\lambda = 0.4$. To design the array, we have to null the pattern in the above directions. If there is no other constraint, we use the following as array factor:

$$AF_d(u) = 1 \tag{4-76}$$

By combining Equations (4-75) and (4-76), we receive the pattern presented in Fig. 4.42. It is obvious that the pattern follows the desired constraints.

If the desired pattern is a Chebyshev polynomial $T_7(x)$ with nulls at $0°$, $60°$ and SLL $= -20$ dB, the pattern observed is given in Fig. 4.43. Obviously, it is shown that the pattern follows the constraints with a small increase in SLL.

Finally, an example of a Taylor pattern is given with two inner side lobes of -40 dB and two outer ones of -20 dB. Also, it is desirable to have nulls at $0°$ and $50°$. In Fig. 4.44, the resulting pattern is presented. It is obvious that nulls follow the constraints but the SLLs increase.

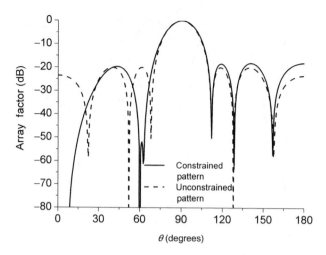

Figure 4.43 Pattern $T_7(x)$ of a linear array of $N = 8$ elements and $d/\lambda = 0.4$ with nulls at $0°$, $60°$ and SLL $= -20$ dB

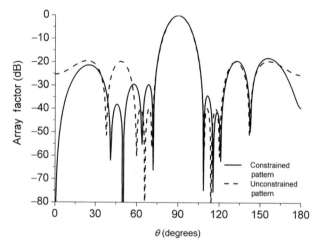

Figure 4.44 Taylor pattern of a linear array of $N = 13$ elements and $d/\lambda = 0.4$ with nulls at $0°$, $60°$ inner SLL $= -40$ dB and outer SLL $= -20$ dB

4.4.2 Synthesis Subject to Minimization of the Pattern Level in Certain Angular Regions

If the constraints have to do with minimization of the pattern level in M different angular regions, the constraint functions are

$$f_n(\ldots, B_i^*, .., B_j, \ldots) = \sum_{m=1}^{M} \frac{\partial P_m}{\partial B_i^*}, i = 1, \ldots, N$$

$$P_m = \int_{u_{m-1}}^{u_m} \left| \sum_{n=1}^{N} B_n \psi_n(u) \right|^2 du$$

(4-77)

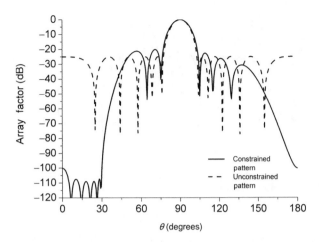

Figure 4.45 A Dolph-Chebyshev array with SLL $= -25$ dB, $N = 11, d/\lambda = 0.5$ and minimum of the pattern in the region $0° \le \theta \le 30°$

From Equation (4-77), we have

$$f_n(\ldots, B_i^*, \ldots, B_j, \ldots) = \sum_{m=1}^{M} \frac{\partial P_m}{\partial B_i^*} = 0, i = 1, \ldots, N$$

$$\sum_{m=1}^{M} \sum_{i=1}^{N} B_i S_{ij}^m = 0, j = 1, \ldots, N \tag{4-78}$$

$$S_{ij}^m = \int_{u_{m-1}}^{u_m} \psi_i(u) \psi_j^*(u) \, du$$

By combining Equations (4-73) and (4-78), we have

$$B_n = \Gamma_n - \sum_{i=1}^{M} \sum_{j=1}^{N} \lambda_j (S_{nj}^i)^* \tag{4-79}$$

Lagrange multipliers are found by substituting B_n of Equation (4-79) to the second equation of Equation (4-78).

An example of a Dolph-Chebyshev linear array with SLL $= -25$ dB, $N = 11$ and $d/\lambda = 0.5$ is given next. It is desirable to have a minimum of the pattern in the region $0° \le \theta \le 30°$. In Fig. 4.45, the resulting pattern is presented.

It is obvious that the pattern follows the constraint but the SLL increases to -20 dB. In the same array, one can have null constraints at the same time by combining the methods given above. In Fig. 4.46, the pattern for an additional constraint with null in $\theta = 120°$ is given.

4.4.3 Synthesis Subject to Constraints on SLLs

If it is desirable to design an array, subject to constraints on the resulting SLLs, two different conditions must be met. One condition defines local maxima and the other the

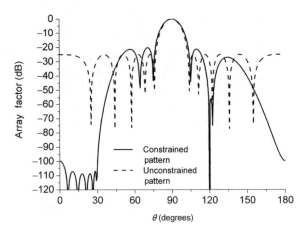

Figure 4.46   A Dolph-Chebyshev array with SLL $= -25$ dB, $N = 11, d/\lambda = 0.5$. Minimum of the pattern is in the region $0° \le \theta \le 30°$ and null in $\theta = 120°$

levels of the maxima. So, for M side lobes, it is

$$\frac{\partial AF(u)}{\partial u}|_{u=u_m} = 0, m = 1, \ldots, M$$

$$AF(u)|_{u=u_m} = T_m$$

(4-80)

or

$$\sum_{n=1}^{N} B_n \frac{\partial \psi_n(u)}{\partial u}|_{u=u_m} = 0 \Rightarrow f_m(B_1, B_2, \ldots B_N) = \sum_{n=1}^{N} B_n S_{nm} = 0, m = 1, \ldots, M$$

(4-81)

$$\sum_{n=1}^{N} B_n \psi_n(u_m) = T_m \Rightarrow\Rightarrow f_{m+M}(B_1, B_2, \ldots B_N) = \sum_{n=1}^{N} B_n \psi_n(u_m) - T_m = 0$$

By combining Equations (4-73) and (4-81), we have

$$B_n = \Gamma_n + \sum_{m=1}^{M} [\lambda_m S_{nm}^* + \lambda_{m+M} \psi_n^*(u_m)]$$

$$\sum_{j=1}^{M} \lambda_j \left(\sum_{n=1}^{N} S_{nj} S_{nm}^* \right) + \sum_{j=1}^{M} \lambda_{j+M} \left(\sum_{n=1}^{N} S_{nj} \psi_n^*(u_m) \right)$$

$$= \sum_{n=1}^{N} \Gamma_n^* S_{nm}^*, m = 1, \ldots, M$$

(4-82)

$$\sum_{j=1}^{M} \lambda_j \left(\sum_{n=1}^{N} \psi_n(u_j) S_{nm}^* \right) + \sum_{j=1}^{M} \lambda_{j+m} \left(\sum_{n=1}^{N} \psi_n(u_j) \psi_n^*(u_m) \right)$$

$$= \sum_{n=1}^{N} \Gamma_n^* \psi_n^*(u_m) - T_m, m = 1, \ldots, M$$

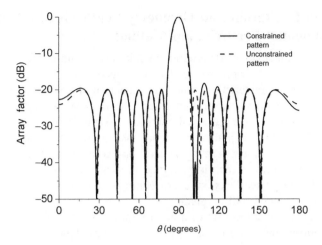

Figure 4.47 Dolph-Chebyshev pattern of a linear array with $N = 15$ elements $d/\lambda = 0.45$, SLL $= -20$ dB and the inner side lobe at $98.5°$ being at -40 dB level

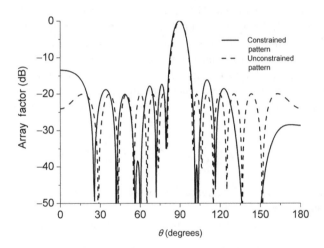

Figure 4.48 Dolph-Chebyshev pattern of a linear array with $N = 15$ elements, $d/\lambda = 0.45$, SLL $= -20$ dB, null at $60°$, minimum in the region $140° - 145°$ and the inner side lobe at $98.5°$, with -40 dB level

The last two equations of Equation (4-82) are solved for λ_j. The excitations B_n are found from the first equation of Equation (4-82). It is noticed that Γ_n is the excitation amplitude given by the orthogonal method without constraints.

Figure 4.47 presents a Dolph-Chebyshev pattern of a linear array with $N = 15$ elements, $d/\lambda = 0.45$ and SLL $= -20$ dB with the additional constraint that the inner side lobe at $98.5°$ will be SLL $= -40$ dB. It is obvious that the pattern follows the constraint. Again, one can combine all the above-mentioned constraints in one array at the same time.

The Dolph-Chebyshev pattern of the same array with null at $60°$, minimum in the region $140°-145°$ and the inner side lobe at $98.5°$, with -40 dB level, is given in Fig. 4.48.

4.5 Quantized Excitation and Geometry Synthesis of a Linear Array: The Orthogonal Perturbation Method

Most of the synthesis techniques determine the unknown excitation for a given geometry and only a few have to do with the design of the geometry of an array. In geometry design, the common characteristic is the adjustment of the spacing among the array elements according to certain optimization criteria. Moffet [58], Ismail and Dawoud [59], Ng [60] and Ng et al. [61] have presented interesting examples of the geometry synthesis. Moreover, recently, in [38, 44], the orthogonal method for the synthesis of uniformly excited arrays has been given. The basic idea has come from the re-forming of the geometry of an initial array, which is perturbed in such a way so as to approximate the desired pattern. In the present section, a procedure that combines an iterative technique with the orthogonal method will be given.

The initial array can be a uniform or a non-uniform one. It is obvious that, in general, the right non-uniform initial array can give better results than those of a uniform one. We must also say that, in practice, it is preferable to use quantized amplitude excitations. Quantization is appropriate in arrays of solid-state modules with saturated output amplifiers. The array can be arranged by certain quantized amplitudes, which come from the corresponding continuous ones. The number of quantized amplitudes depends on the desired pattern and could be up to the number of the array elements. One can find several interesting works in synthesis of arrays with quantized weights in the literature. Most of them use statistical thinning to control the pattern side lobes. The above studies, with typical examples in [62, 63], do not specifically address the pattern approximation, and so their patterns are different than the ones that are given with the orthogonal method.

Suppose that we have an N-element linear array (see Fig. 4.49) with the array factor expressed by

$$AF_0(\theta) = \sum_{i=1}^{N} I_i e^{j\beta d_i \sin\theta} = \sum_{i=1}^{N} I_i \phi_i(\theta) \tag{4-83}$$

We can perturb the pattern by changing the position d_i of each element. The perturbation δd_i is made in such a way that $\beta(\delta d_i) \ll 1$.

In this case, the exponential term of Equation (4-83) can be approximated to

$$e^{j\beta(d_i + \delta d_i)\sin\theta} \cong [1 + j\beta(\delta d_i)\sin\theta] e^{j\beta d_i \sin\theta} \tag{4-84}$$

With the help of Equation (4-84), the array factor becomes

$$AF_1(\theta) \cong \sum_{i=1}^{N} [1 + j\beta(\delta d_i)\sin\theta] I_i e^{j\beta d_i \sin\theta} \tag{4-85}$$

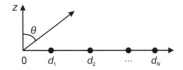

Figure 4.49   Geometry of an N-element linear array

By subtracting Equation (4-83) from Equation (4-85) and by dividing $\sin\theta$, we have

$$AF(\theta) = \frac{AF_1(\theta) - AF_0(\theta)}{\sin\theta} = \sum_{i=1}^{N} A_i\phi_i(\theta) \tag{4-86}$$

where

$$A_i = j\beta(\delta d_i)I_i \tag{4-87}$$

It must be clarified that for $\theta = 0°$ and $180°$, $AF(\theta)$ is already kept equal to zero. Also, it is obvious that $AF(\theta)$ has a similar expression to $AF_0(\theta)$. In both factors, the determination of I_i and A_i can be achieved by applying the orthogonal method. By following the orthogonalization procedure, we can establish an orthogonal basis $\{\psi_i(\theta)\}$ coming from $\{\phi_i(\theta)\}$.

It is

$$\psi_i(\theta) = \sum_{j=1}^{i} C_j^{(i)}\phi_j(\theta) \tag{4-88}$$

In the new basis, the array factors are expressed in the form

$$\left.\begin{array}{l} AF_0(\theta) = \sum_{i=1}^{N} B_i^0\psi_i(\theta) \\[2mm] AF(\theta) = \sum_{i=1}^{N} B_i\psi_i(\theta) \end{array}\right\} \tag{4-89}$$

The amplitudes B_i^0 and B_i are found from the inner products as follows.

$$\left.\begin{array}{l} B_i^o = <AF^o(\theta), \psi_i(\theta)> \\ B_i = <AF(\theta), \psi_i(\theta)> \end{array}\right\} \tag{4-90}$$

Finally, I_i and A_i are related to the above by

$$I_i = \sum_{j=i}^{N} B_j^o C_i^{(j)} \tag{4-91}$$

$$A_i = \sum_{j=i}^{N} B_j C_i^{(j)} \tag{4-92}$$

Equation (4-91) gives the element excitation for the desired pattern $AF_0(\theta)$. If, instead of I_i, we use quantized approximate values, an error to the resultant pattern will occur. The error depends on the approximation error. It must be pointed out that quantization of the amplitudes offers easier implementation of the arrays. As noticed above, arrays consisting of solid-state modules with saturated output amplifiers excite quantized amplitudes. Of course, the above arrays give patterns that do not fit exactly but approximate the desired ones.

An array with quantized amplitudes can be used as the initial one. The array is perturbed and the amplitudes A_i give δd_i. Under the perturbation assumption, $AF_1(\theta) - AF_0(\theta)$ must be kept relatively small. Therefore, a significant change of the initial pattern $AF_0(\theta)$ can be obtained only by using an iterative procedure. The final position of the array elements is found in the last iteration, where the suitable approximation of the desired pattern has been obtained.

It is noticed that the above procedure does not give a successful outcome for any number of quantized amplitudes. The success of the whole issue comes from a suitable selection of the initial array.

The procedure must be repeated for larger numbers of quantized amplitudes if the final result does not give the desired approximation.

4.5.1 Detailed Analysis and Examples

Suppose that for a given array geometry and a desired pattern, we have the element excitation that can be found by using the orthogonal method. Our first step is the quantization procedure. It must be pointed out that one can use several algorithms. Suppose that for an array with N elements, it is desirable to have L quantized amplitudes ($L \leq N$). We first normalize the amplitudes by dividing all of them with the minimum one.

If at least two of the normalized amplitudes are very close to each other in a margin around 0.5, then we multiply all of them by a factor that has the initial value of 2. Otherwise, their values remain unchanged.

The quantized values are the roundup integers of the above amplitudes. If their number is less than L, then we increase the multiplication factor by 1 and the procedure is repeated until at least L quantized values are found.

If L is equal to 1, then, no further calculations are needed and all the quantized amplitudes are assigned the value of 1. If L is equal to 2, then the two quantized amplitudes are the maximum ones and the mean integer value of all the other values.

If L is more than 2 then, the next value is found in the same way by calculating the mean integer value of all the other amplitudes.

Each one of the amplitudes is approximated by the integer that is closest to its real value from the set of integer values found. If more than one integer value is equally close to the amplitude, then the larger integer is selected. The whole process is shown in the flow chart of Fig. 4.50.

A design example of an array with 13 elements and a desired array factor $AF(\theta) = e^{-(a\pi \sin\theta)^2} \cos(2b\pi \sin\theta)$ is given next. The desired pattern has a maximum at $\theta = 0°$, SLL $= -22$ dB and null at $\theta = 12°$. The coefficients are $a = 1.557$ and $b = 1.202$. Three quantized amplitudes are chosen. Figure 4.51 shows the desired and the computed patterns, while Table 4.5 gives the positions and the amplitudes of the elements of the array. It is obvious that the solution with the above quantized amplitudes is acceptable.

A 12-element array with the Chebyshev polynomial $T_{11}(x)$ as array factor and SLL $= -20$ dB can be designed by using only one type of amplitude. Figure 4.52 shows the pattern and Table 4.6 presents the positions of the elements.

Table 4.5 Array geometry and amplitudes for $L = 3$ of the pattern of Fig. 4.51

n	1,13	2,12	3,11	4,10	5,9	6,8	7
d_n/λ	±3.381	±3.167	±2.524	±1.798	±1.140	±0.581	0.000
A_n	2	2	9	16	16	16	16

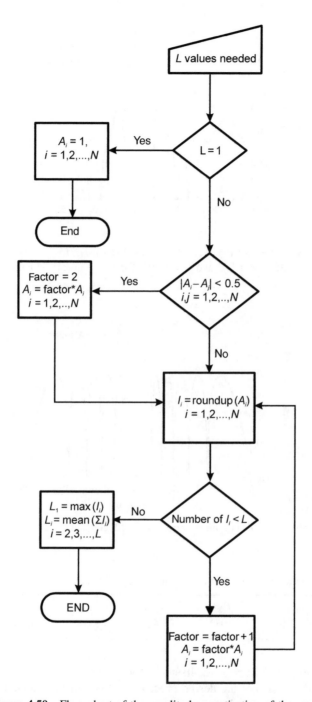

Figure 4.50 Flow chart of the amplitude quantization of the array

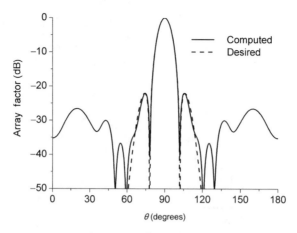

Figure 4.51 Array factor, $AF(\theta) = e^{-(a\pi \sin\theta)^2} \cos(2b\pi \sin\theta)$, for 13 elements with null at $\theta = 12°$, SLL $= -22$ dB, $a = 1.557$ and $b = 1.202$, for $L = 3$ quantized amplitudes

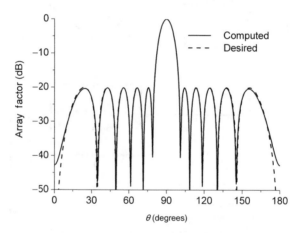

Figure 4.52 Pattern of a $T_{11}(x)$ linear array with 12 elements, SLL $= -20$ dB and uniform excitation

Table 4.6 Array geometry of the pattern of Fig. 4.52

n	1,12	2,11	3,10	4,9	5,8	6,7
d_n/λ	±2.740	±2.082	±1.481	±1.046	±0.570	±0.231

The same array with SLL $= -25$ dB needs four quantized amplitudes. Figure 4.53 and Table 4.7 give the details of the design procedure. It is understandable that the increased number of quantized amplitudes comes from the type of constraints of the array factor.

It is noticed that the same pattern can be realized with only two non-quantized amplitudes. Table 4.7 gives the geometry and the excitation of the quantized and the non-quantized arrays.

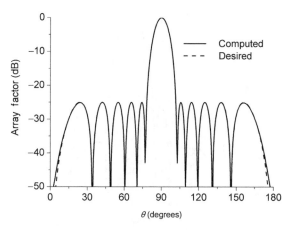

Figure 4.53 Pattern of a $T_{11}(x)$ linear array with 12 elements, SLL $= -25$ dB and $L = 4$ quantized amplitudes

Table 4.7 Geometry and amplitudes of two arrays. One with $L = 4$ quantized amplitudes and the other with two non-quantized amplitudes of the pattern of Fig. 4.53

	n	1,12	2,11	3,10	4,9	5,8	6,7
Quantized	d_n/λ	±2.750	±2.260	±1.781	±1.250	±0.760	±0.274
	A_n	2	2	3	4	4	5
Non-quantized	d_n/λ	±2.750	±2.158	±1.778	±1.268	±0.708	±0.242
	A_n	0.506	0.506	0.506	1.000	1.000	1.000

Table 4.8 Array geometry and amplitudes for $L = 3$ of the pattern given in Fig. 4.54

n	1,8	2,7	3,6	4,5
d_n/λ	±1.801	±1.222	±0.678	±0.258
A_n	2	4	4	5

Table 4.9 Geometries of the arrays with the patterns of Fig. 4.55

$N = 13$	n	1,13	2,12	3,11	4,10	5,9	6,8	7
	d_n/λ	±3.443	±2.732	±2.027	±1.445	±0.946	±0.461	0.000
$N = 14$	n	1,14	2,13	3,12	4,11	5,10	6,9	7,8
	d_n/λ	±3.440	±2.813	±2.132	±1.592	±1.091	±0.663	±0.198

A $T_4(x)$ desired array factor with SLL $= -25$ dB and HPBW $= 25°$ can be designed with three quantized amplitudes and eight elements. The results are shown in Fig. 4.54 and Table 4.8.

For a $T_{14}(x)$ with SLL $= -20$ dB, if we choose an array with 13 or 14 uniformly excited, instead of 15 non-uniformly excited, elements, we receive acceptable results with SLL ≤ -20 dB. Figure 4.55 and Table 4.9 give the details.

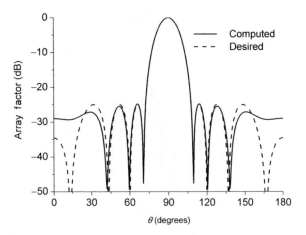

Figure 4.54 Pattern $T_4(x)$ with SLL $= -25$ dB and HPBW $= 25°$ of a linear array with eight elements

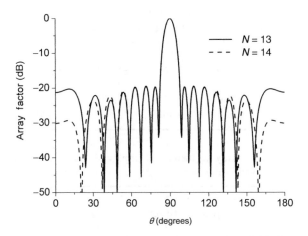

Figure 4.55 Patterns of uniformly excited arrays with $N = 13$ and 14 elements for $T_{14}(x)$ with SLL $= -20$ dB

Table 4.10 Geometry of the array with the pattern given in Fig. 4.56

n	1,31	2,30	3,29	4,28	5,27	6,26	7,25	8,24
d_n/λ	±7.144	±7.046	±6.263	±5.520	±4.883	±4.320	±3.802	±3.307
n	9,23	10,22	11,21	12,20	13,19	14,18	15,17	16
d_n/λ	±2.869	±2.392	±2.027	±1.536	±1.226	±0.744	±0.414	0.000

Another interesting example of an array with 31 uniformly excited elements is given next. It is desirable to have a pattern $T_{30}(x)$ with HPBW $= 3.6°$ and SLL $\cong -20$ dB. Figure 4.56 and Table 4.10 give the pattern and the geometry of the array.

As a final example, the array pattern behaviour versus frequency is presented. A $T_{15}(x)$ array factor for SLL $= -25$ dB, 16 elements and two quantized amplitudes is given.

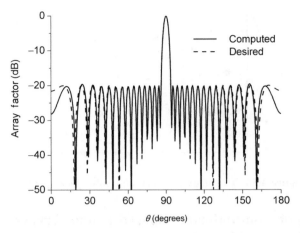

Figure 4.56 Pattern $T_{30}(x)$ with HPBW $= 3.6°$ and SLL $\cong -20$ dB of a uniformly excited array with 31 elements

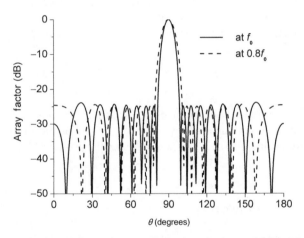

Figure 4.57 Pattern of 16 elements and two quantized excitations with $T_{15}(x)$ desired factor and SLL $= -25$ dB. The pattern is given for f_0 and $0.8f_0$

Table 4.11 Array geometry and excitations for $L = 2$ of the pattern of Fig. 4.57

n	1,16	2,15	3,14	4,13	5,12	6,11	7,10	8,9
d_n/λ	± 3.760	± 3.385	± 2.766	± 2.633	± 1.973	± 1.328	± 0.779	± 0.252
A_n	2	2	2	2	5	5	5	5

Figure 4.57 shows the pattern at a design frequency f_0 and at a frequency $0.8f_0$. The pattern at both frequencies keeps SLL close to -25 dB, while the HPBW increases from $7.4°$ to about $9.265°$. Obviously, it is $9.265° \times 0.8 \approx 7.4°$. Table 4.11 gives the array characteristics.

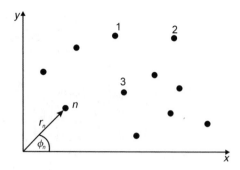

Figure 4.58   Geometry of a non-uniformly spaced planar array

4.6 Synthesis of Non-uniformly Spaced Planar Arrays: The Orthogonal Method

The synthesis of non-uniformly spaced planar arrays for a given geometry and a desired array factor on the same plane can be obtained by using the orthogonal method. Let us consider first a non-uniformly spaced planar array with N elements (see Fig. 4.58).

The array factor $AF(\phi)$ on the same plane is

$$AF(\phi) = \sum_{n=1}^{N} I_n e^{j\beta r_n \cos(\phi - \phi_n)} \tag{4-93}$$

where r_n, ϕ_n are the cylindrical coordinates of the array elements.

In general, the basis functions $\phi_i(\phi) = e^{j\beta r_i \cos(\phi - \phi_i)}$ of Equation (4-93) are not orthogonal. With the well-known procedure, an orthogonal basis $\{\psi_i(\phi)\}$ can be derived. It is

$$\psi_1(\phi) = \frac{\phi_1(\phi)}{\langle \phi_1(\phi), \phi_1(\phi) \rangle^{\frac{1}{2}}} \tag{4-94}$$

$$\psi_n(\phi) = \frac{\phi_n(\phi) - \sum_{j=1}^{n-1} \langle \phi_n(\phi), \psi_j(\phi) \rangle \psi_j(\phi)}{\left\{ \int_0^{2\pi} \left[\phi_n(\phi) - \sum_{j=1}^{n-1} \langle \phi_n(\phi), \psi_j(\phi) \rangle \psi_j(\phi) \right]^2 d\phi \right\}^{\frac{1}{2}}} \tag{4-95}$$

$\psi_n(\phi)$ is the orthonormalized function expressed by the equation

$$\psi_n(\phi) = \sum_{i=1}^{n} C_i^{(n)} \phi_i(\phi) \tag{4-96}$$

The symbol $\langle \phi_n(\phi), \psi_j(\phi) \rangle$ represents the inner product given by

$$\langle \phi_n(\phi), \psi_j(\phi) \rangle = \int_0^{2\pi} \phi_n(\phi) \psi_j^*(\phi) \, d\phi \tag{4-97}$$

The coefficients $C_j^{(n)}$ can be found from Equations (4-95) to (4-97) in a manner similar to that in Equation (4-9):

$$
\left.
\begin{aligned}
C_n^{(n)} &= \frac{1}{D_n} \\
C_k^{(n)} &= -\frac{2\pi}{D_n} \sum_{j=k}^{n-1} \sum_{i=1}^{j} C_k^{(j)} C_i^{(j)} k_{in} \\
D_n &= \left\{ 2\pi - 4\pi^2 \sum_{j=1}^{n-1} \left(\sum_{i=1}^{j} C_i^{(j)} k_{in} \right)^2 \right\}^{1/2} \\
k_{in} &= 2\pi J_0(\beta e_{in}) \\
e_{in} &= (r_n \sin\phi_n - r_i \sin\phi_i) / \sin\phi_{ni} \\
\cot\phi_{ni} &= (r_n \cos\phi_n - r_i \cos\phi_i)/(r_n \sin\phi_n - r_i \sin\phi_i)
\end{aligned}
\right\}
\qquad (4\text{-}98)
$$

(J_0 is the Bessel function of zero order)

The difference between Equations (4-9) and (4-98) is in the expression of k_{in}.
Equation (4-93) is modified to

$$
AF(\phi) = \sum_{i=1}^{N} B_i \psi_i(\phi)
\qquad (4\text{-}99)
$$

where

$$
B_i = \langle AF(\phi), \psi_i(\phi) \rangle
\qquad (4\text{-}100)
$$

Finally, the excitations I_i result as follows:

$$
I_i = \sum_{j=i}^{N} B_j C_i^{(j)}
\qquad (4\text{-}101)
$$

From the above procedure, it is shown that for a given non-uniformly planar array, the element excitations can be achieved. The accuracy of the approximation depends on the desired array factor, the number and the position of the array elements. The whole synthesis follows the five steps given in Section 4.3.

We start our examples with a circular array of $N = 16$ uniformly positioned elements. The radius of the array is $r = 1\lambda$ and the desired array factor is $\delta(0°)$, which gives maximum directivity. Figure 4.59 presents the pattern of the array. It is observed that HPBW $= 18.38°$ and SLL $\cong -13.5$ dB.

The pattern of the same array with a Dolph-Chebyshev desired array factor is presented in Fig. 4.60. It is noticed that the projection of the array at the axis normal to the maximum direction shows nine elements. Thus, the $T_8(x)$ is used with a mean equal distance of $d/\lambda = 0.25$.

The pattern of the same array with a Chebyshev (case 1) desired factor is presented in Fig. 4.61. Here, a $T_4(x)$ is used with the same mean distance as above.

If, in the Chebyshev pattern, the HPBW is desired, case 4 is used. Figure 4.62 presents the pattern for SLL $= -25$ dB and HPBW $= 30°$.

An exponential array factor of the form $AF(\phi) = e^{-a\pi \sin\phi}$ that gives a desired HPBW depending on a is presented in Fig. 4.63. It is obvious that the orthogonal method can give the desired pattern.

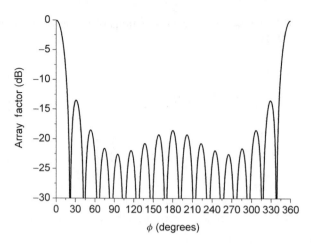

Figure 4.59 Pattern of a circular array with uniformly positioned elements. It is $AF(\phi) = \delta(0°)$, $N = 16$ and $r = 1\lambda$

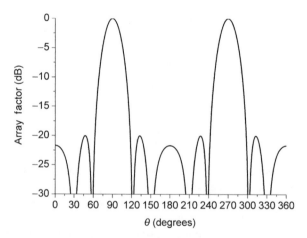

Figure 4.60 Dolph-Chebyshev pattern of a circular array with uniformly positioned elements. It is $AF(\phi) = T_8(x)$, SLL $= -20$ dB, $N = 16$ and $r = 1\lambda$

It is noticed that the synthesis of the planar arrays in the present form is suitable for elements with omni patterns on their plane.

4.7 Synthesis of Non-uniformly Spaced 3-D Arrays: The Orthogonal Method

The synthesis of general (3-D) non-uniformly spaced arrays for a given geometry and a desired array factor is possible by using the orthogonal method. Let us consider first an array of N discrete, similarly oriented, identical elements arranged arbitrarily in a 3-D space (see Fig. 4.64).

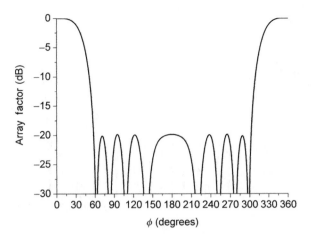

Figure 4.61 Chebyshev (case 1) pattern of a circular array with uniformly positioned elements. It is $AF(\phi) = T_4(x)$, SLL $= -20$ dB, $N = 16$ and $r = 1\lambda$

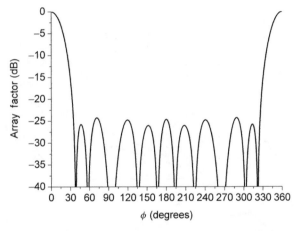

Figure 4.62 Chebyshev (case 4) pattern of a circular array with uniformly positioned elements. It is $AF(\phi) = T_4(x)$, SLL $= -25$ dB, HPBW $= 30°$, $N = 16$, and $r = 1\lambda$

The radiation pattern of such an array is given by

$$F(\theta, \phi) = g(\theta, \phi) \sum_{n=1}^{N} I_n e^{j\beta r_n \cos \xi_n} \qquad (4\text{-}102)$$

where $\cos \xi_n = \sin \theta \sin \theta_n \cos(\phi - \phi_n) + \cos \theta \cos \theta_n$ and (r_n, θ_n, ϕ_n) are the spherical coordinates of a convenient reference point of the nth element. The term $g(\theta, \phi)$ represents the element radiation pattern. Equation (4-102) is transformed into

$$AF(\theta, \phi) = F(\theta, \phi)/g(\theta, \phi) = \sum_{n=1}^{N} I_n e^{j\beta r_n \cos \xi_n} = \sum_{n=1}^{N} I_n \phi_n(\theta, \phi) \qquad (4\text{-}103)$$

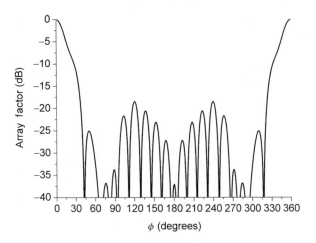

Figure 4.63 Pattern of a circular array with uniformly positioned elements. It is, $AF(\phi) = e^{-a\pi \sin\phi}$, $a = 7.261$, HPBW $= 25°$, $N = 16$ and $r = 1.5\lambda$

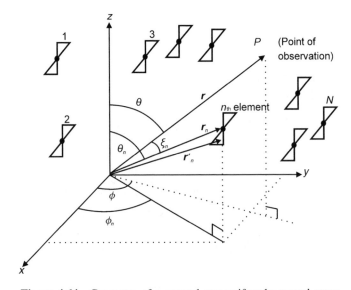

Figure 4.64 Geometry of a general non-uniformly spaced array

The currents I_n can be derived in the case where the array factor $AF(\theta, \phi)$ is given.

In a non-uniformly spaced linear and/or planar array, the array factor is a function of one independent variable. Here, $AF(\theta, \phi)$ depends on both θ and ϕ. Before we start the orthogonal procedure, it is important to prove that the infinite set of complex functions

$$\phi_1(\theta, \phi), \phi_2(\theta, \phi), \ldots, \phi_n(\theta, \phi), \ldots\ldots$$

is linearly independent. This is important because it proves that the synthesis problem is not ill conditioned because of the basis of the vector space. A set of elements in a vector space is linearly independent if none of the elements in the set can be written as a linear combination of finitely many other vectors in the set. For our set of complex functions, it is enough to prove that the Gramian [26] of the set is different than zero. The Gramian, known also as 'Gram determinant', comes from the Gramian matrix. This is a real symmetric matrix that can be used to test the linear independence of functions. In our case, the Gramian is

$$
G = \begin{vmatrix}
\langle \phi_1, \phi_1 \rangle & \langle \phi_1, \phi_2 \rangle & \cdots & \langle \phi_1, \phi_N \rangle \\
\langle \phi_2, \phi_1 \rangle & \langle \phi_2, \phi_2 \rangle & \cdots & \langle \phi_2, \phi_N \rangle \\
\vdots & \vdots & \ddots & \vdots \\
\langle \phi_N, \phi_1 \rangle & \langle \phi_N, \phi_2 \rangle & \cdots & \langle \phi_N, \phi_N \rangle
\end{vmatrix}
\tag{4-104}
$$

Symbol $\langle \phi_i, \phi_j \rangle$ represents the inner product given by

$$
\langle \phi_i, \phi_j \rangle = \int_0^\pi \int_0^{2\pi} \phi_i \phi_j^* \sin \theta \, d\phi \, d\theta
\tag{4-105}
$$

By solving Equation (4-105), we can find that [26] $\langle \phi_i, \phi_j \rangle$ is

$$
\langle \phi_i, \phi_j \rangle = 4\pi \frac{\sin \beta r_{ij}}{\beta r_{ij}} = 4\pi k_{ij}
\tag{4-106}
$$

where $r_{ij} = |r_i - r_j|$.

The Gramian becomes

$$
G = (4\pi)^N \begin{vmatrix}
1 & k_{12} & k_{13} & \cdots & k_{1N} \\
k_{21} & 1 & k_{23} & \cdots & k_{2N} \\
. & . & . & \cdots & . \\
k_{N1} & k_{N2} & . & \cdots & 1
\end{vmatrix}
\tag{4-107}
$$

In the non-uniform linear arrays, $\langle \phi_i, \phi_j \rangle$ is of the same form as Equation (4-106). The only difference is in the constant 4π that becomes 2π. So, we see that every array is of the same Gramian form.

In a linear array, the functions $\phi_i(u) = e^{jx_i u}$ are linearly independent. This can be proven.

If $\phi_1(u), \phi_2(u), \ldots, \phi_n(u), \ldots\ldots$ are linearly dependent, then

$$
\begin{vmatrix}
\displaystyle\sum_{i=1}^N \lambda_i \phi_i(u) = 0 \\[2mm]
\displaystyle\sum_{i=1}^N \lambda_i \frac{d\phi_i(u)}{du} = 0 \\[2mm]
\displaystyle\sum_{i=1}^N \lambda_i \frac{d^2\phi_i(u)}{du^2} = 0 \\[2mm]
\cdots\cdots\cdots\cdots\cdots \\[2mm]
\displaystyle\sum_{i=1}^N \lambda_i \frac{d^{N-1}\phi_i(u)}{du^{N-1}} = 0
\end{vmatrix}
\tag{4-108}
$$

It is $\lambda_i \neq 0$.

From Equation (4-108), it must be that

$$\begin{vmatrix} \phi_1 & \phi_2 & \phi_3 & \cdots & \phi_N \\ \dfrac{d\phi_1}{du} & \dfrac{d\phi_2}{du} & \dfrac{d\phi_3}{du} & \cdots & \dfrac{d\phi_N}{du} \\ \vdots & \vdots & \vdots & \cdots & \vdots \\ \dfrac{d^{N-1}\phi_1}{(du)^{N-1}} & \vdots & \vdots & \cdots & \vdots \end{vmatrix} = 0 \tag{4-109}$$

Equation (4-109) is written as

$$\exp[j(x_1 + x_2 + \cdots + x_N)u] \begin{vmatrix} 1 & 1 & \cdots & 1 \\ x_1 & x_2 & \cdots & x_N \\ \vdots & \vdots & \cdots & \vdots \\ x_1^{N-1} & x_2^{N-1} & & x_N^{N-1} \end{vmatrix} = 0 \tag{4-110}$$

It is obvious that Equation (4-110) is not true because the determinant is a Vandermonde one [26], which, for $x_1 \neq x_2 \neq \ldots\ldots \neq x_N$, is different from zero.

So, the functions $\phi_1(u), \phi_2(u), \ldots, \phi_n(u), \ldots\ldots$ as it is proven, are linearly independent and their Gramian is not zero. Since the Gramian is the same as Equation (4-107), we deduce that $\phi_1(\theta, \phi), \phi_2(\theta, \phi), \ldots, \phi_n(\theta, \phi), \ldots\ldots$ are also linearly independent.

The basis functions $\phi_i(\theta, \phi)$ can be orthonormalized with the aid of

$$\psi_n(\theta,) = \sum_{i=1}^{n} C_i^{(n)} \phi_i(\theta, \phi) \tag{4-111}$$

where the coefficients are

$$\left. \begin{aligned} C_i^{(n)} &= \frac{1}{D_n} \\ C_k^{(n)} &= -\frac{4\pi}{D_n} \sum_{j=k}^{n-1} \sum_{i=1}^{j} C_k^{(j)} C_i^{(j)} k_{in} \\ D_n &= \left\{ 4\pi - 4^2\pi^2 \sum_{j=1}^{n-1} \left(\sum_{i=1}^{j} C_i^{(j)} k_{in} \right)^2 \right\}^{1/2} \\ k_{in} &= \frac{\sin \beta r_{in}}{\beta r_{in}} \end{aligned} \right\} \tag{4-112}$$

By following the same procedure as in linear and planar arrays, we have

$$B_i = \langle AF(\theta, \phi), \psi_i(\theta, \phi) \rangle \tag{4-113}$$

$$I_i = \sum_{j=i}^{N} B_j C_i^{(j)} \tag{4-114}$$

The above-proven relations are general and can be applied to any array—linear, planar or 3-D. So, a general non-uniformly spaced array can be synthesized in a quite simplified way.

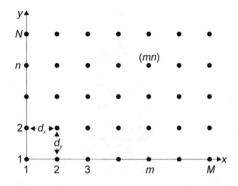

Figure 4.65 The geometry of a rectangular array

4.7.1 Detailed Analysis and Examples

Individual radiators positioned on a line can be synthesized in the same way as a linear array. For a planar array, where it is desirable to have the synthesis in 3-D space, there are two independent variables ϕ, θ and it is obvious that the present procedure must be applied. The most usual planar arrays are the rectangular ones. The elements are positioned along a rectangular grid. The grid can be made from the combination of two perpendicular linear arrays. Because of the increase of the variables, the pattern can be controlled and scanned at any point in space. The more symmetrical patterns and the lower side lobes are the two main advantages of planar over linear arrays. Let us refer to Fig. 4.65.

The element corresponding to the mth row and the nth column is the mnth element with excitation I_{mn}. The array factor is of the form

$$AF(\theta, \phi) = \sum_{m=0}^{M-1} \sum_{n=0}^{N-1} I_{mn} z_x^m z_y^n \tag{4-115}$$

where

$$z_x = e^{j\beta d_x \sin\theta \cos\phi} \quad \text{and} \quad z_y = e^{j\beta d_y \sin\theta \sin\phi}$$

A rectangular array can be the product of two different non-uniform linear arrays. The excitations are $I_{mn} = I_m I_n$, where $I_{m/n}$ is the current of the $m^{\text{th}}/n^{\text{th}}$ element of the linear array at the x/y-axis. By using two Dolph-Chebyshev arrays with SLL $= -20$ dB for 8×6 elements, we have the 3-D pattern shown in Fig. 4.66 for $d_x = d_y = \lambda/2$. It is observed that the maximum is oriented at $\theta_0 = 0$ and $\theta_0 = \pi$, and the side lobes are at -20 dB level.

Let us assume that N isotropic elements are spaced on a circular ring of radius R (see Fig. 4.67).

The array factor is written in the following form:

$$AF(\theta, \phi) = \sum_{n=1}^{N} I_n e^{j[\beta R \sin\theta \cos(\phi - \phi_n)]} = \sum_{n=1}^{N} I_n \phi_n(\theta, \phi) = \sum_{n=1}^{N} B_n \psi_n(\theta, \phi) \tag{4-116}$$

To have a pattern with maximum directivity at $\theta = 0°, \phi = 0°$ it must be

$$AF(\theta, \phi) = \delta(0, 0) \tag{4-117}$$

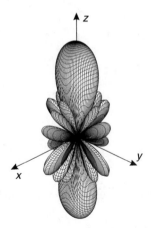

Figure 4.66   Three-dimensional pattern of an 8×6 planar Chebyshev array with SLL $= -20$ dB, $d_x = d_y = 0.5\lambda$

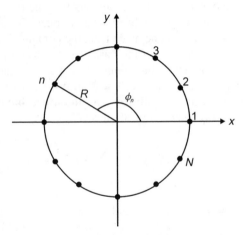

Figure 4.67   Geometry of a circular array with N elements

By using Equations (4-111)-(4-114), the currents of the array can be found from B_n:

$$B_n = \psi_n^*(0, 0) \tag{4-118}$$

The pattern of a 12-element circular array with $\beta R = 10$ is shown in Fig. 4.68. It is noticed that the present case is different than the case given in Section 4.6, where the desired maximum is on the plane of the array.

An interesting case of a circular array is that coming from a corner reflector. A dipole positioned at the bisector of a corner reflector with angle $\omega_N = \pi/N$ produces $2N - 1$ images. By adding up the contributions from the dipole and its images, we can derive the total radiation pattern. Figure 4.69 presents a $\pi/3$ corner reflector with its images. It is obvious that the array is a circular one. It is noticed that the real pattern is the one in front of the corner.

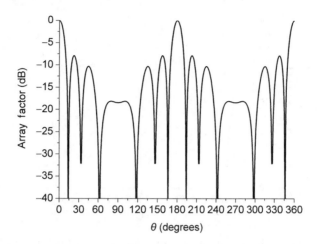

Figure 4.68 Planar pattern at $\phi = 0°$ of a circular array with $AF(\theta, \phi) = \delta(0, 0)$, $N = 12$ and $\beta R = 10$

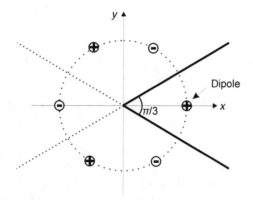

Figure 4.69 A vertical dipole with its images in front of a $\pi/3$ corner reflector

M circular arrays in concentric rings produce an array factor of the form

$$AF(\theta, \phi) = \sum_{m=1}^{M} \sum_{n=1}^{N} I_{mn} e^{j[\beta R_m \sin\theta \cos(\phi - \phi_{mn})]} \qquad (4\text{-}119)$$

I_{mn} is the excitation of the nth element of the mth ring.

A usual case of concentric rings is that of a corner reflector with a linear array positioned in front of it. For a seven parallel dipole linear array in front of a $\pi/2$ corner reflector, the total pattern comes from Equation (4-119) with $M = 7$ and $N = 4$. By taking the images into account, we calculate the excitations as above. If the linear array is a Chebyshev (case 4) with SLL $= -20$ dB and HPBW $= 31°$, the total pattern, which is given in Fig. 4.70, is taken from Equation (4-119).

Another interesting array is a cylindrical one. Parallel circular arrays with their centres on the same axis perform a cylindrical array. This is a special kind of a 3-D array. The

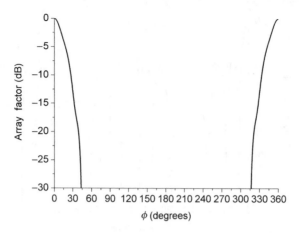

Figure 4.70   Radiation pattern of a seven-dipole linear array in front of a $\pi/2$ corner reflector. The distance between the elements and the corner is $d = \lambda/4$. The pattern of the linear array is $T_3(x)$ with SLL $= -20$ dB and HPBW $= 31°$

array factor is written as

$$AF(\theta, \phi) = \sum_{m=1}^{M} \sum_{l=1}^{L} I_{ml} e^{j[\beta R_0 \sin\theta \cos(\phi - \phi_{ml}) + \beta z_m \cos\theta]} \tag{4-120}$$

I_{ml} is the excitation of the lth element of the mth circular array.

R_0 is the radius of the circular arrays and z_m is the position of the mth array on the z-axis.

A linear array in front of a corner with its axis parallel to the edge is a virtual cylindrical array. Such an array is extremely useful for applications in radio and mobile communications. The excitation can be uniform or not depending on the application. Figure 4.71 presents an array of collinear dipoles in front of a $\pi/2$ corner reflector.

To have a maximum at $\theta = 90°, \phi = 45°$ without other constraints, the array should be uniform. If there are additional constraints on the SLL, then the excitations can be non-uniform on the z-axis. Figure 4.72 shows the pattern taken by using the orthogonal method of an array with SLL ≤ -20 dB and maximum at $\theta = 90°, \phi = 45°$.

4.8 Synthesis of Non-uniformly Spaced 3-D Arrays with Arbitrarily Oriented Dipoles: The Non-parallel Orthogonal Method

The advantage of the methods of synthesizing antenna arrays having an arbitrary distribution of their elements is that they allow the designer more degrees of freedom. It is obvious that maximum freedom is achieved when no restriction regarding either the location or the orientation of the elements is imposed. In this section, the design of an array consisting of not only arbitrarily positioned but also arbitrarily oriented dipoles will be presented.

Let us consider first an array of N discrete, arbitrarily spaced and oriented identical dipoles (see Fig. 4.73).

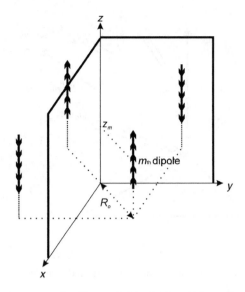

Figure 4.71   An array of collinear dipoles in front of a $\pi/2$ corner reflector

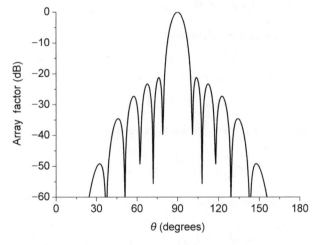

Figure 4.72   Vertical pattern at $\phi = 45°$ of a Chebyshev linear array in front of a $\pi/2$ corner reflector with $N = 9$ collinear dipoles, at distance $d = 0.7\lambda$ and SLL ≤ 20 dB

To specify a dipole, besides the three coordinates of location, two more coordinates are needed. These are the directional coordinates that will specify the radiation properties of each dipole. Thus, the nth dipole is assigned r_n, θ_n, ϕ_n as coordinates of location or, alternatively, x_n, y_n, z_n and θ^n, ϕ^n as coordinates of direction.

The intensity of the electric field of the array is

$$E(\theta, \phi) = \sum_{n=1}^{N} E_n(\theta, \phi) = \sum_{n=1}^{N} I_n \boldsymbol{\phi}_n(\theta, \phi) \qquad (4\text{-}121)$$

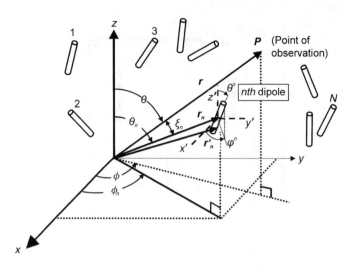

Figure 4.73   Geometry of a general non-uniformly spaced array of arbitrarily oriented dipoles

$E_n(\theta, \phi)$ is the electric field of the nth dipole and $\boldsymbol{\phi}_n(\theta, \phi)$ is the corresponding 'function vector' of the normalized electric field. It is

$$\boldsymbol{\phi}_n(\theta, \phi) = [Z_n(\theta, \phi)\hat{\boldsymbol{\theta}} - H_n(\theta, \phi)\hat{\boldsymbol{\phi}}]e^{j\beta r_n \cos \xi_n} \tag{4-122}$$

where

$$Z_n(\theta, \phi) = \cos \theta \sin \theta^n \cos(\phi - \phi^n) - \sin \theta \cos \theta^n$$
$$H_n(\theta, \phi) = \sin \theta^n \sin(\phi - \phi^n) \tag{4-123}$$

We define the inner product of two 'function vectors' as

$$\langle \boldsymbol{\phi}_i, \boldsymbol{\phi}_j \rangle = \int_0^\pi \int_0^{2\pi} \boldsymbol{\phi}_i \cdot \boldsymbol{\phi}_j^* \sin \theta \, d\phi \, d\theta \tag{4-124}$$

From Equation (4-124), we get [28, 32]

$$\langle \boldsymbol{\phi}_i, \boldsymbol{\phi}_j \rangle = \frac{8\pi}{3} k_{ij} = (B_{ij} + D_{ij}) \sin \theta^i \sin \theta^j - A_{ij}^j \cos \theta^i \sin \theta^j$$
$$- A_{ij}^i \cos \theta^j \sin \theta^i + G_{ij} \cos \theta^i \cos \theta^j \tag{4-125}$$

where

$$B_{ij} = \frac{3}{4}\cos(\phi^i - \phi^j)\sqrt{\frac{\pi}{2}}\left\{ \frac{J_{1/2}(\beta r_{ij})}{(\beta r_{ij})^{1/2}} - 2\frac{J_{3/2}(\beta r_{ij})}{(\beta r_{ij})^{3/2}} + (\beta R_{ij})^2\frac{J_{5/2}(\beta r_{ij})}{(\beta r_{ij})^{5/2}} \right\}$$
$$- \frac{3}{4}\cos(2\phi_{ij} + \phi^i - \phi^j)\sqrt{\frac{\pi}{2}} \cdot \left\{ 2\sqrt{\frac{\pi}{2}}\frac{J_{1/2}[\beta(r_{ij} + z_{ij})/2]J_{1/2}[\beta(r_{ij} - z_{ij})/2]}{(\beta r_{ij})} \right.$$
$$\left. - \frac{J_{1/2}(\beta r_{ij})}{(\beta r_{ij})^{1/2}} - (\beta R_{ij})^2\frac{J_{5/2}(\beta r_{ij})}{(\beta r_{ij})^{5/2}} \right\}, i \neq j$$

$$B_{ij} = \frac{3}{4}, i = j \tag{4-126}$$

$$D_{ij} = \frac{3}{4}\cos(\phi^i - \phi^j)\sqrt{\frac{\pi}{2}}\left\{\frac{J_{1/2}(\beta r_{ij})}{(\beta r_{ij})^{1/2}}\right\} - \frac{3}{4}\cos(2\phi_{ij} + \phi^i - \phi^j)$$

$$\sqrt{\frac{\pi}{2}}\left\{2\sqrt{\frac{\pi}{2}}\frac{J_{1/2}[\beta(r_{ij} + z_{ij})/2]J_{1/2}[\beta(r_{ij} - z_{ij})/2]}{(\beta r_{ij})} - \frac{J_{1/2}(\beta r_{ij})}{(\beta r_{ij})^{1/2}}\right\}, i \neq j$$

$$D_{ij} = \frac{3}{4}, i = j \tag{4-127}$$

$$A_{ij}^j = -\frac{3}{2}\cos(\phi_{ij} - \phi^j)\sqrt{\frac{\pi}{2}}(\beta^2 z_{ij} R_{ij})\frac{J_{5/2}(\beta r_{ij})}{(\beta r_{ij})^{5/2}}, i \neq j$$

$$A_{ij}^j = 0, i = j \tag{4-128}$$

$$G_{ij} = \frac{3}{2}\sqrt{\frac{\pi}{2}}\left\{2\frac{J_{3/2}(\beta r_{ij})}{(\beta r_{ij})^{3/2}} - (\beta R_{ij})^2\frac{J_{5/2}(\beta r_{ij})}{(\beta r_{ij})^{5/2}}\right\}, i \neq j$$

$$G_{ij} = 1, i = j \tag{4-129}$$

The symbols of the above equations are as follows:

$$
\begin{aligned}
r_{ij} &= \left[(x_i - x_j)^2 + (y_i - y_j)^2 + (z_i - z_j)^2\right]^{1/2} \\
R_{ij} &= \left[(x_i - x_j)^2 + (y_i - y_j)^2\right]^{1/2} \\
z_{ij} &= z_i - z_j \\
\phi_{ij} &= \phi_i - \phi_j
\end{aligned}
\tag{4-130}
$$

With the aid of Equation (4-122), it is possible to construct 'function vectors' related to each other by a relation analogous to that of orthogonality. We follow the Gram-Schmidt procedure and we get

$$\psi_n(\theta, \phi) = \sum_{i=1}^{n} C_i^{(n)}\phi_i(\theta, \phi) \tag{4-131}$$

where

$$
\left.
\begin{aligned}
C_n^{(n)} &= \frac{1}{D_n} \\
C_k^{(n)} &= -\frac{8\pi/3}{D_n}\sum_{j=k}^{n-1}\sum_{i=1}^{j} C_k^{(j)}C_i^{(j)}k_{in} \\
D_n &= \left\{\frac{8\pi}{3} - \left(\frac{8\pi}{3}\right)^2\sum_{j=1}^{n-1}\left(\sum_{i=1}^{j} C_i^{(j)}k_{in}\right)^2\right\}^{1/2}
\end{aligned}
\right\}
\tag{4-132}
$$

By following a similar procedure to that of the earlier sections we have

$$L_i = \langle E(\theta, \phi), \psi_i(\theta, \phi)\rangle \tag{4-133}$$

$$I_i = \sum_{j=i}^{N} L_j C_i^{(j)} \tag{4-134}$$

As a result of the above technique, one can say that it is possible to synthesize an antenna consisting of arbitrarily oriented dipoles. The synthesis follows the steps given above. The only difference is that the determination of the position of the dipoles is also supplemented by their direction.

As an example, consider the case of two dipoles normal to each other.

We start with the definition of the five coordinates of each dipole:

$$r_1 = \theta_1 = \phi_1 = 0, \phi^1 = 0, \theta^1 = \frac{\pi}{2}$$

$$r_2 = \theta_2 = \phi_2 = 0, \phi^2 = \frac{\pi}{2}, \theta^2 = \frac{\pi}{2} \tag{4-135}$$

The coefficients $C_j^{(n)}$ are

$$C_1^{(1)} = \frac{1}{\left[\dfrac{8\pi}{3}\right]^{1/2}}, C_1^{(2)} = 0, C_2^{(2)} = \frac{1}{\left[\dfrac{8\pi}{3}\right]^{1/2}} \tag{4-136}$$

The desired electric field is assumed to be

$$\boldsymbol{E}(\theta, \phi) = e^{j\phi}\hat{\boldsymbol{\phi}}_0 \tag{4-137}$$

The quantities L_1, L_2 are found to be the following:

$$L_1 = 2\pi, L_2 = -2j\pi \tag{4-138}$$

Finally, the excitation currents are

$$I_1 = 1, I_2 = 1e^{-j\pi/2} \tag{4-139}$$

The above values give the expected well-known omnidirectional element [64].

Another example involves an antenna array that provides maximum directivity. The magnitude of the electric field must be an impulse in the direction from which the emission is sought. $\boldsymbol{E}(\theta, \phi)$ can thus be expressed as

$$\boldsymbol{E}(\theta, \phi) = \delta(\theta - \theta_0)\delta(\phi - \phi_0)(K\hat{\boldsymbol{\theta}} + \Lambda\hat{\boldsymbol{\phi}}) \tag{4-140}$$

K and Λ are, in general, complex quantities defining the polarization.

From Equations (4-133) and (4-140), we have

$$L_i = (K\hat{\boldsymbol{\theta}} + \Lambda\hat{\boldsymbol{\phi}}) \cdot \boldsymbol{\psi}_i^*(\theta_0, \phi_0) \tag{4-141}$$

Obviously, maximum directivity is

$$D = \frac{|\boldsymbol{E}(\theta_0, \phi_0)|^2}{\dfrac{1}{4\pi} \int_0^{2\pi} \int_0^{\pi} |\boldsymbol{E}(\theta, \phi)|^2 \sin\theta \, d\theta \, d\phi} = 4\pi \frac{\left|\sum\limits_{i=1}^{N} L_i \boldsymbol{\psi}_i(\theta_0, \phi_0)\right|^2}{\sum\limits_{i=1}^{N} L_i L_i^*} \tag{4-142}$$

$\psi_i(\theta, \phi)$ can be written as

$$\boldsymbol{\psi}_i(\theta, \phi) = \psi_i{}'(\theta, \phi)\hat{\theta} + \psi_i{}''(\theta, \phi)\hat{\phi} \tag{4-143}$$

From Equations (4-141) and (4-142), D turns out to be

$$D = 4\pi \frac{\left|\sum_{i=1}^{N} L_i \psi_i{}'(\theta_0, \phi_0)\right|^2 + \left|\sum_{i=1}^{N} L_i \psi_i{}''(\theta_0, \phi_0)\right|^2}{\sum_{i=1}^{N} L_i L_i^*} \tag{4-144}$$

Equation (4-144) can be expressed in the form

$$D = D_\theta + D_\phi \tag{4-145}$$

where

$$D_\theta = 4\pi \frac{\left|\sum_{i=1}^{N} L_i \psi_i{}'(\theta_0, \phi_0)\right|^2}{\sum_{i=1}^{N} L_i L_i^*} \tag{4-146}$$

and

$$D_\phi = 4\pi \frac{\left|\sum_{i=1}^{N} L_i \psi_i{}''(\theta_0, \phi_0)\right|^2}{\sum_{i=1}^{N} L_i L_i^*} \tag{4-147}$$

By using the Schwartz inequality, we have

$$D_\theta = 4\pi \frac{\left|\sum_{i=1}^{N} L_i \psi_i{}'(\theta_0, \phi_0)\right|^2}{\sum_{i=1}^{N} L_i L_i^*} \leq 4\pi \frac{\left(\sum_{i=1}^{N} L_i L_i^*\right)\left(\sum_{i=1}^{N} \psi_i{}'\psi_i{}'^*\right)}{\sum_{i=1}^{N} L_i L_i^*} = 4\pi \sum_{i=1}^{N} \psi_i{}'\psi_i{}'^* \tag{4-148}$$

$$D_\phi = 4\pi \frac{\left|\sum_{i=1}^{N} L_i \psi_i{}''(\theta_0, \phi_0)\right|^2}{\sum_{i=1}^{N} L_i L_i^*} \leq 4\pi \frac{\left(\sum_{i=1}^{N} L_i L_i^*\right)\left(\sum_{i=1}^{N} \psi_i{}''\psi_i{}''^*\right)}{\sum_{i=1}^{N} L_i L_i^*} = 4\pi \sum_{i=1}^{N} \psi_i{}''\psi_i{}''^* \tag{4-149}$$

From Equations (4-145), (4-148) and (4-149), it is

$$D = D_\theta + D_\phi \leq 4\pi \sum_{i=1}^{N} (\psi_i{}'\psi_i{}'^* + \psi_i{}''\psi_i{}''^*) = 4\pi \sum_{i=1}^{N} \boldsymbol{\psi}_i(\theta_0, \phi_0) \cdot \boldsymbol{\psi}_i^*(\theta_0, \phi_0) \tag{4-150}$$

From Equation (4-150), maximum directivity becomes

$$D_{\max} = 4\pi \sum_{i=1}^{N} \boldsymbol{\psi}_i(\theta_0, \phi_0) \cdot \boldsymbol{\psi}_i^*(\theta_0, \phi_0) \tag{4-151}$$

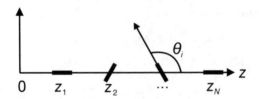

Figure 4.74   A linear array with non-parallel dipoles in equal distance where $\theta^i = \dfrac{i-1}{N-1}\pi$, $\phi^i = 0$

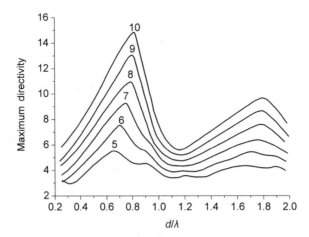

Figure 4.75   Maximum directivity at $\theta = 90°$, $\phi = 90°$ of the array of Fig. 4.74 versus d/λ, for $N = 5-10$ dipoles

A linear array with non-parallel dipoles on the z-axis (see Fig. 4.74), where $\theta^i = \dfrac{i-1}{N-1}\pi$ and $\phi^i = 0$, has maximum directivity at $\theta = 90°$, $\phi = 90°$ versus d/λ for $N = 5-10$ dipoles, which is given in Fig. 4.75.

Maximum directivity at $\theta = 90°$, $\phi = 90°$ of a similar array with the dipoles directed at $\theta^i = \dfrac{(i-1)}{(N-1)}\dfrac{\pi}{2}$, $\phi^i = 0$ is presented in Fig. 4.76.

Both figures show that there is a certain distance d/λ where maximum directivity takes its optimum value.

It is obvious that one can follow a procedure similar to the one given above for arrays with elements other than dipoles.

4.9 Synthesis of Arrays of Wire Antennas: The MoM Orthogonal Method

The synthesis of arrays consisting of non-parallel wire antennas is a generalization of the synthesis given in the last section. Here, suppose that we have a wire structure (Fig. 4.77) with N_1 straight segments.

By the well-defined method of moments (MoM) [65, 66], the wire structure can be analysed in a straightforward fashion using a matrix manipulation. The currents and the

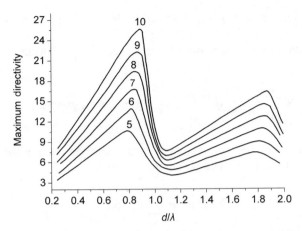

Figure 4.76 Maximum directivity at $\theta = 90°, \phi = 90°$ of a linear array with non-parallel dipoles at an equal distance, where $\theta^i = \dfrac{(i-1)}{(N-1)} \dfrac{\pi}{2}, \phi^i = 0$, versus d/λ, for $N = 5-10$ dipoles

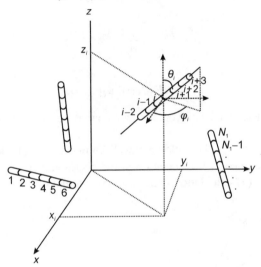

Figure 4.77 Geometry of a wire structure

voltages of the segments are related by

$$[I] = [\Upsilon][V] \tag{4-152}$$

If we have N voltage generators at the $N(N \leq N_1)$ from the N_1 segments, then the total electric field of the structure is

$$E(\theta, \phi) = \sum_{n=1}^{N_1} I_n \left[E_n^\theta(\theta, \phi) + E_n^\phi(\theta, \phi) \right]$$

$$= \sum_{i=1}^{N} V_i \sum_{i=1}^{N_1} \Upsilon_{ni} \left[E_n^\theta(\theta, \phi) + E_n^\phi(\theta, \phi) \right] = \sum_{i=1}^{N} V_i \boldsymbol{\phi}_i(\theta, \phi) \tag{4-153}$$

I_n is the nth element of $[I]$, Υ_{ni} is the nith element of $[\Upsilon]$ and V_i is the ith feed voltage.

We define the inner product of two 'function vectors' in a manner similar to that in Equation (4-124):

$$\langle \boldsymbol{\phi}_i, \boldsymbol{\phi}_j \rangle = \sum_{n=1}^{N_1} \sum_{m=1}^{N_1} \left[\Upsilon_{ni} \Upsilon_{mj}^* \int_0^\pi \int_0^{2\pi} (E_n^\theta \cdot E_m^{\theta*} + E_n^\phi \cdot E_m^{\phi*}) \sin\theta \, d\phi \, d\theta \right] = \frac{8\pi}{3} k_{ij}$$
(4-154)

We construct 'function vectors' related to each other by a relation analogous to that of orthogonality by following the Gram-Schmidt procedure:

$$\boldsymbol{\psi}_n(\theta, \phi) = \sum_{i=1}^n C_i^{(n)} \boldsymbol{\phi}_i(\theta, \phi)$$
(4-155)

The coefficients $C_i^{(n)}$ are given by Equation (4-132).

Now, the unknown values of the structure are the feed voltages, which are

$$V_i = \sum_{j=i}^N L_j C_i^{(j)}$$
(4-156)

where

$$L_i = <E(\theta, \phi), \boldsymbol{\psi}_i(\theta, \phi)>$$
(4-157)

It is obvious that if $E(\theta, \phi)$ is given, we can synthesize the antenna by computing the feed voltages.

The synthesis implements the following well-known steps: (1) definition of the five coordinates of each segment, (2) calculation of the coefficients $C_i^{(j)}$, (3) evaluation of the electric field $E(\theta, \phi)$, (4) calculation of L_i and (5) calculation of the input voltages.

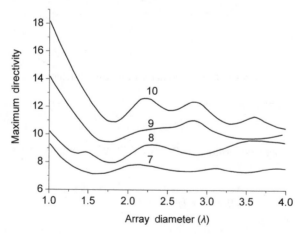

Figure 4.78 Maximum directivity of a circular array with tangential $\lambda/2$ wire antennas in directions $\theta^i = \dfrac{\pi}{4}$, $\phi^i = \dfrac{\pi}{2} + \dfrac{2\pi(i-1)}{N}$, for $N = 7-10$, versus the diameter of the array

An example of the maximum directivity of a circular array with tangential $\lambda/2$ wire antennas in directions $\theta^i = \dfrac{\pi}{4}, \phi^i = \dfrac{\pi}{2} + \dfrac{2\pi(i-1)}{N}$, for $N = 7{-}10$, versus the diameter of the array, is presented in Fig. 4.78.

Another interesting case is a linear array of parallel wire $\lambda/2$ dipoles. It is desirable to have a Chebyshev array with SLL $= -20$ dB. In Table 4.12, the required input currents and voltages for five dipoles are given.

The resulting patterns for both spacings are shown in Fig. 4.79. If the mutual coupling is not taken into account, then, the relative input voltages are the same as the currents. In this case, the real patterns (uncorrected) are different than the desired ones and are also given in Fig. 4.79.

From Figure 4.79, one can see that, for $d/\lambda = 0.5$, the desired pattern is closer to the uncorrected one. This is not the case for $d/\lambda = 0.25$, where the mutual coupling is strong and the side lobe becomes -7.22 dB. By considering the uncorrected patterns, one can see that, for small distances among the elements of the array, the difference between the uncorrected and the desired ones is large. This means that it is almost impossible for the same feed voltages in point source and wire antenna array to produce the same patterns.

An example of a planar array with six $\lambda/2$ dipoles is given. The array is presented in Fig. 4.80. The desired pattern contains nulls at $\theta = 90°$ and $\phi = 110°, 135°, 180°, 300°$.

Table 4.12 Relative input currents and voltages for a five-element Chebyshev array with SLL $= -20$ dB and equal spacing $d/\lambda = 0.25$ & 0.5

Element No	$d/\lambda = 0.25$		$d/\lambda = 0.5$	
	I_i	V_i	I_i	V_i
1,5	1	1	1	1
2,4	-1.194	$1.90225\lfloor 1.56°$	1.608	$1.383\lfloor 341.2°$
3	2.178	$3.027\lfloor 32.09°$	1.932	$1.804\lfloor -5.97°$

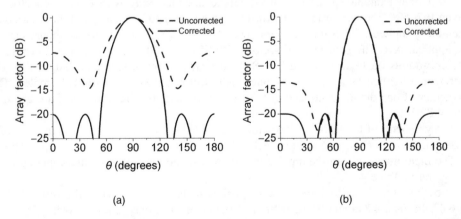

(a)   (b)

Figure 4.79 Corrected and uncorrected patterns of a five-element linear array with -20 dB and inter-element distance d; (a) 0.25λ (b) 0.5λ

Table 4.13 Required input currents and voltages of the array of Fig. 4.80

Element number	1	2	3	4	5	6
$I_i(dB)$	0.0|0.0°	8.13|−88.5°	4.12|177.2°	0.0|147.2°	8.13|−124.3°	4.12|−30.1°
$V_i(dB)$	0.0|0.0°	16.13|161.74°	23.105|58.63°	25.51|31.08°	22.46|121.22°	2.58|221.44°

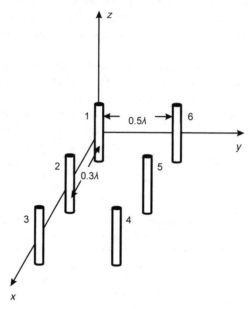

Figure 4.80 Geometry of a planar array with six (3×2) parallel $\lambda/2$ dipoles

The required input currents and voltages are given in Table 4.13, and the pattern is given in Fig. 4.81.

As a final example, a mast-mounted dipole antenna array with corner reflectors is given [67, 68]. The wire-grid model for the antenna structure is used, and the orthogonal method is applied. The corner reflectors are of 120° each. Thus, the image theory cannot be applied. A 17-dB gain in each sector, with SLL $\leq$ −20 dB, is desirable. Because of the constraints on high gain and low SLL, the array in each sector is designed by the orthogonal method as a Chebyshev one. Each sector contains nine folded dipoles. The geometry of one dipole positioned in front of a mast and its wire-grid model are given in Fig. 4.82.

The geometry of the three sectors with nine folded collinear dipoles, each in front of a common mast, is presented in Fig. 4.83.

The input impedance of the dipoles in each sector and the corresponding input currents are given in Table 4.14.

In each sector, the HPBW on the horizontal plane is 120° and that on the vertical plane is 6.4°. In Figs. 4.84a and b, the pattern of the antenna in one sector is given.

If it is desirable to have a HPBW of 60° on the horizontal plane, then the whole geometry of the reflectors is changed. Figure 4.85 shows the new geometry.

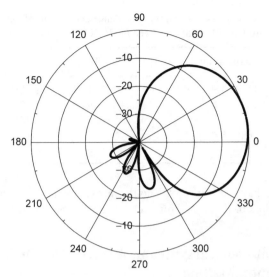

Figure 4.81 Pattern of the array of Fig. 4.80 with desired nulls at $\theta = 90°$ and $\phi = 110°$, $135°$, $180°$, $300°$

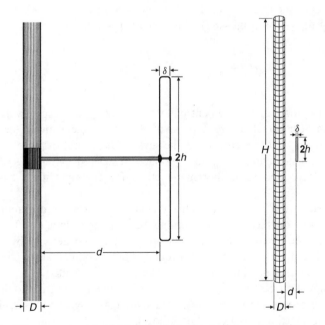

Figure 4.82 The geometry of a folded dipole in front of a mast and its wire-grid model

The input impedance of the dipoles in each sector and the corresponding input currents are given in Table 4.15.

In each sector, the HPBW on the horizontal plane is now $60°$ and on the vertical plane it is $6.4°$. In Figs. 4.86a and b, the pattern of the antenna in one sector is given.

Table 4.14   The input impedance and current of the dipoles of the array (given in Fig. 4.83) in one sector

Element number	Zin (Ω)	I_{in} (A)
1,9	$411.59 + j2.81$	1.7287
2,8	$373.43 - j9.49$	1.7687
3,7	$381.25 - j7.28$	2.3342
4,6	$382.68 - j5.96$	2.7313
5	$383.00 - j5.61$	2.8742

If it is desired to have an array with uniform excitation, the orthogonal perturbation method could be applied. Examples for such a case can be found in [67, 68].

4.10 Synthesis of General Antenna Arrays: The Orthogonal Compensation Method

The analysis of an array is relatively simple when the coupling between the elements is not taken into account. In such a case, the radiation pattern is found as the product of the element pattern and the array factor. Let us suppose a linear array with parallel-uncoupled elements along the z-axis (Fig. 4.87). The pattern of the antenna array can be expressed as follows.

$$AF(\theta, \phi) = f^i(\theta, \phi) \cdot \sum_{n=1}^{N} I_n^d e^{j\beta z_n \cos\theta} \qquad (4\text{-}158)$$

where $f^i(\theta, \phi)$ is the isolated element pattern, I_n^d and z_n are respectively the excitation current and the position of the feed point of the nth element along the z-axis, and N is the total number of the elements of the array. The synthesis problem of the above antenna involves calculation of I_n^d that produces a desired pattern $AF(\theta, \phi)$. This is a straightforward task, and the orthogonal method for dealing with it has already been presented.

In a real system, the pattern of each element of an array is different than $f^i(\theta, \phi)$ because it depends on the existent coupling between the elements. It is obvious that the radiation properties of a real antenna array cannot be derived by the uncoupling consideration. The divergence between a real and an uncoupled array appears as a noise floor on the pattern that precludes low side lobes or certain pattern nulls. The divergence is evident in signal-processing arrays, which are extremely sensitive to small errors due to the non-linear processing procedures. Under the coupling conditions, the pattern of a linear antenna array can be expressed as follows.

$$AF(\theta, \phi) = \sum_{n=1}^{N} I_n f_n(\theta, \phi) e^{j\beta z_n \cos\theta} \qquad (4\text{-}159)$$

where I_n is the input current and $f_n(\theta, \phi)$ is the pattern of the nth element. This pattern is actually the antenna pattern when only the nth element is excited and the rest are

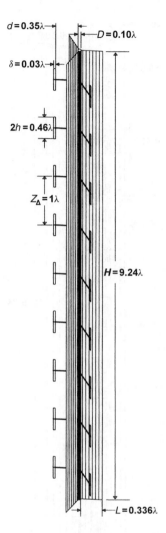

$d=0.35\lambda$
$D=0.10\lambda$
$\delta=0.03\lambda$
$2h=0.46\lambda$
$Z_\Delta=1\lambda$
$H=9.24\lambda$
$L=0.336\lambda$

Figure 4.83 Geometry of a three-sector collinear dipole array with HPBW of 20° on the horizontal plane

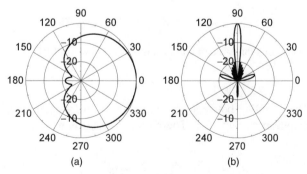

Figure 4.84 The (a) horizontal and (b) vertical pattern of one sector of the antenna of Fig. 4.83

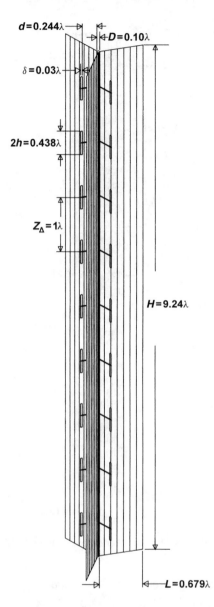

Figure 4.85 Geometry of a three-sector collinear dipole array with HPBW of 60° on the horizontal plane

short-circuited. The coupling between the mth and the nth element can be presented by the equivalent excitation current C_{nm}. Therefore, $f_n(\theta, \phi)$ is written as

$$f_n(\theta, \phi) = \sum_{m=1}^{N} C_{nm} f^i(\theta, \phi) e^{j\beta(z_m - z_n)\cos\theta} \qquad (4\text{-}160)$$

Table 4.15 The input impedance and current of the dipoles of the array (given in Fig. 4.85) in one sector

Element number	Zin (Ω)	I_{in} (A)
1, 9	$349.94 + j12.82$	1.7287
2, 8	$324.44 - j5.21$	1.7687
3, 7	$330.65 - j1.87$	2.3342
4, 6	$331.47 - j0.32$	2.7313
5	$331.60 - j0.08$	2.8742

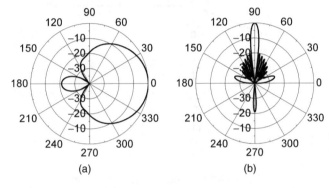

(a)   (b)

Figure 4.86 The (a) horizontal and (b) vertical pattern of one sector of the antenna of Fig. 4.85

The exponential term reflects the phase difference produced by the relative position between the mth and the nth element.

After substituting Equation (4-160) to Equation (4-159) and doing some algebra, we get

$$AF(\theta, \phi) = f^i(\theta, \phi) \cdot \sum_{n=1}^{N} \left(\sum_{m=1}^{N} C_{mn} I_m \right) \cdot e^{j\beta z_n \cos\theta} \qquad (4\text{-}161)$$

From Equations (4-158) and (4-161), it is found that

$$I_n^d = \sum_{m=1}^{N} C_{mn} I_m \qquad (4\text{-}162)$$

Equation (4-162) can be written in matrix form as follows.

$$[I^d] = [C]^T \cdot [I] \qquad (4\text{-}163)$$

where $[]^T$ indicates the matrix transpose. $[I^d]$ is the excitation of the uncoupled array, while $[I]$ is the corresponding excitation for the real array. By solving Equation (4-163), we have

$$[I] = \{[C]^T\}^{-1} \cdot [I^d] \qquad (4\text{-}164)$$

A technique to find C_{nm} has been suggested in the past [69, 70]. It is based on the application of the Fourier analysis on the pattern $f_n(\theta, \phi)$ given in Equation (4-160).

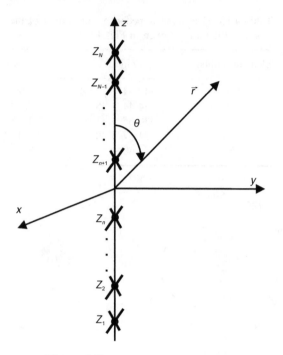

Figure 4.87 Antenna array geometry

When the distance between the elements is less than $\lambda/2$, the accuracy of the calculation of the coupling coefficients becomes a major issue.

The coupling coefficients, irrespective of the inter-element distance, can be derived by the orthogonal compensation method that is given next [25–36, 47]. In Equation (4-160), $f_n(\theta, \phi)$ is expressed as a linear combination of non-orthogonal functions $\phi_m^{(n)}(\theta, \phi)$, where

$$\phi_m^{(n)}(\theta, \phi) = f^i(\theta, \phi)e^{j\beta(z_m - z_n)\cos\theta} \qquad (4\text{-}165)$$

Thus,

$$f^n(\theta, \phi) = \sum_{m=1}^{N} C_{nm}\phi_m^{(n)}(\theta, \phi) \qquad (4\text{-}166)$$

Since the basis functions $\phi_m^{(n)}(\theta, \phi)$ are not orthogonal, a set of orthogonal ones $\psi_i^{(n)}(\theta, \phi)$ can be constructed. These are

$$\psi_i^{(n)}(\theta, \phi) = \sum_{m=1}^{i} C_{m,n}^{(i)}\phi_m^{(n)}(\theta, \phi) \qquad (4\text{-}167)$$

In the new basis, the element pattern $f^n(\theta, \phi)$ is expressed as follows

$$f^n(\theta, \phi) = \sum_{i=1}^{N} B_{ni}\psi_i^{(n)}(\theta, \phi) \qquad (4\text{-}168)$$

The coefficients $C_{m,n}^{(i)}$ can be found from [47]

$$
\left.
\begin{aligned}
C_{1,n}^{(1)} &= \frac{1}{\sqrt{k_{11}^{(n)}}} \\[2mm]
D_i^{(n)} &= \left[k_{ii}^{(n)} - \sum_{j=1}^{i-1} \left(\sum_{\ell=1}^{j} C_{\ell,n}^{(j)} k_{i\ell}^{(n)} \right)^2 \right]^{1/2} , \quad 2 \leq i \leq N \\[2mm]
C_{m,n}^{(i)} &= -\frac{1}{D_i^{(n)}} \sum_{j=m}^{i-1} \left(C_{m,n}^{(j)} \sum_{\ell=1}^{j} C_{\ell,n}^{(j)} k_{i\ell}^{(n)} \right) , \quad m < i \\[2mm]
C_{i,n}^{(i)} &= \frac{1}{D_i^{(n)}}
\end{aligned}
\right\}
\qquad (4\text{-}169)
$$

$k_{i\ell}^{(n)}$ is the inner products of the non-orthogonal basis functions and are defined as follows

$$
k_{i\ell}^{(n)} = \langle \phi_i^{(n)}, \phi_\ell^{(n)} \rangle = \int_0^{2\pi} \int_0^{\pi} \phi_i^{(n)}(\theta,\phi) \phi_\ell^{(n)*}(\theta,\phi) \sin\theta \, d\theta \, d\phi \qquad (4\text{-}170)
$$

Because of the orthogonality of $\{\psi_i^{(n)}(\theta,\phi)\}$, the weights B_{ni} are given by

$$
B_{ni} = \langle f_n, \psi_i^{(n)} \rangle = \int_0^{2\pi} \int_0^{\pi} f_n(\theta,\phi) \psi_i^{(n)*}(\theta,\phi) \sin\theta \, d\theta \, d\phi \qquad (4\text{-}171)
$$

From B_{ni} the coupling coefficients are found as

$$
C_{nm} = \sum_{i=m}^{N} B_{ni} C_{m,n}^{(i)} \qquad (4\text{-}172)
$$

It is obvious that if the element pattern $f_n(\theta,\phi)$ is known, the orthogonal compensation method helps create matrix [C]. $f_n(\theta,\phi)$ can be taken by measuring the element pattern. This pattern, as it was mentioned earlier, is the antenna pattern when only the nth element is excited and the rest are short-circuited. By using Equation (4-164), the excitations can be derived, when coupling coefficients are taken into account.

The above formulation can be extended to 2-D and 3-D arrays as well. In a general array, the exponential factor of Equation (4-159) will be modified to $e^{j\beta(x_n \sin\theta \cos\phi + y_n \sin\theta \sin\phi + z_n \cos\theta)}$, where x_n, y_n, z_n are the coordinates of the centre of the nth element of the array. Equation (4-160) thus becomes

$$
f^n(\theta,\phi) = \sum_{m=1}^{N} C_{nm} f^i(\theta,\phi) e^{j\beta[(x_m - x_n)\sin\theta \cos\phi + (y_m - y_n)\sin\theta \sin\phi + (z_m - z_n)\cos\theta]} \qquad (4\text{-}173)
$$

The procedure of orthogonalization of the basis functions of Equation (4-173) can be found in the literature [26].

In the following text, examples of Chebyshev broadside and end-fire arrays with specified SLLs will be presented. The element pattern, $f^n(\theta,\phi)$, can be derived by the MoM or by measurements [47]. In measurements, the pattern $f^n(\theta, \phi)$ is derived by the signal at the receiving probe when only the nth element is fed and the array is rotated. Usually, in practice, $|f^n(\theta, \phi)|$ is considered, and information about the phase of the pattern is not given. Since it is desirable to have a technique applicable and compatible with measurements, $|f^n(\theta, \phi)|$ is used instead of the complex pattern $f^n(\theta, \phi)$. Consequently, $|f^n(\theta, \phi)|$ for every array element is used and then the orthogonal compensation method is applied on each $|f^n(\theta, \phi)|$ in order to calculate the coupling coefficient matrix.

As a first example, a linear array of five parallel $\lambda/2$ dipoles is examined. The excitation distribution is suitably derived to produce a Chebyshev broadside pattern with SLL $= -20$ dB. The results for three different inter-element distances (0.75λ, 0.50λ and 0.30λ) are shown in Figs. 4.88a, 4.89a and 4.90a respectively. The above figures represent the desired pattern, the pattern without the coupling coefficients, the pattern taken by using

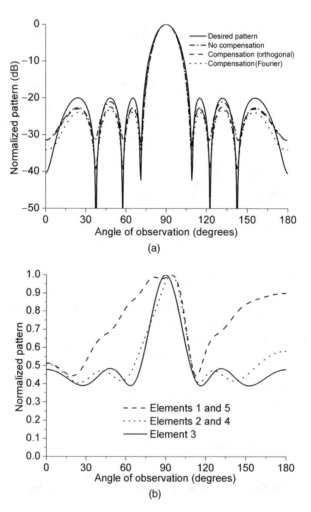

Figure 4.88 Linear array of five dipoles with 0.75λ inter-element distance. (a) Broadside Chebyshev patterns with SLL $= -20$ dB. (b) Patterns of the array elements

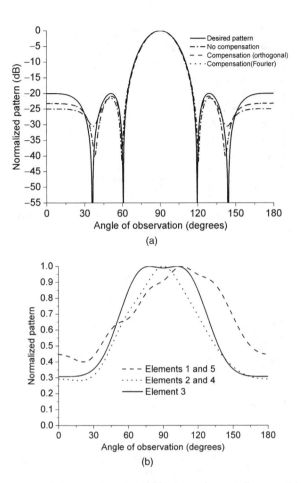

Figure 4.89 Linear array of five dipoles with 0.50λ inter-element distance. (a) Broadside Chebyshev patterns with SLL = −20 dB. (b) Patterns of the array elements

the orthogonal compensation method and the technique based on the Fourier analysis. It is obvious that the patterns change as the inter-element distance decreases. It is obvious that the results taken by the present method are closer to the desired ones. The patterns of the elements of the array, on a plane perpendicular to the dipoles, are shown in Figs. 4.88b, 4.89b and 4.90b. From the above figures, it is obvious that inter-element distance plays an important role. If, instead of $|f^n(\theta, \phi)|$, the $f^n(\theta, \phi)$ was used, the resultant pattern with the present method would be exactly the same as the desired one.

In the second example, an eight-dipole Dolph-Chebyshev broadside array with SLL = −30 dB and inter-element distance equal to 0.517λ is shown. This array is the same as that given in [70]. The results are shown in Fig. 4.91a, while in Fig. 4.91b the patterns of the elements of the array are presented.

Two linear arrays of five and nine parallel λ/2 dipoles with an inter-element distance of 0.20λ are examined next. It is desirable to produce Chebyshev broadside patterns with SLLs equal to −20 dB and −30 dB respectively. The results are presented in Figs. 4.92

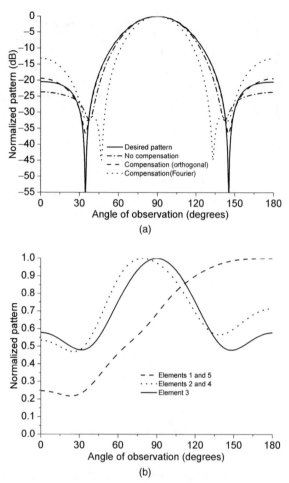

Figure 4.90 Linear array of five dipoles with 0.30λ inter-element distance (a) Broadside Chebyshev patterns with SLL = −20 dB. (b) Patterns of the array elements

and 4.93. It is obvious that the orthogonal compensation method for small inter-element
distance works better than the Fourier method.

Finally, two linear Chebyshev end-fire arrays with five and nine parallel λ/2 dipoles are
studied (see Figs. 4.94 and 4.95). The desired SLLs are −20 dB and −30 dB respectively.
The end-fire patterns are constructed by applying two different cases (cases 2 and 3).

From the above given examples, it is evident that the orthogonal decomposition method
can give acceptable design solutions for the antenna arrays. However, Fig. 4.95b demon-
strates the necessity of considering the phase of the element pattern for tight constraints.

4.11 Synthesis of Conformal Arrays: The Conformal Orthogonal
Method

It is well known that the synthesis problem of conformal arrays is more complex than that
of the corresponding linear and planar ones. The elements of a conformal array do not

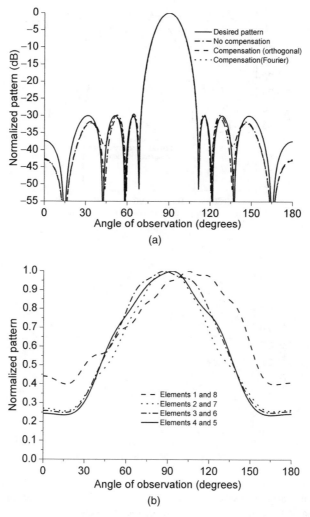

Figure 4.91 Linear array of eight dipoles with inter-element distance equal to 0.517λ. (a) Broadside Chebyshev patterns with SLL = −30 dB. (b) Patterns of the array elements

contribute equally because of the differences in their patterns and polarization. Several approaches available for the synthesis of non-planar arrays could be applied to conformal ones. The projection method given in [71] and the generalized one of Bucci et al. [72, 73] show very promising results. Also, the least mean-square-error (LMSE) method in a matrix inversion form, as well as in the form of the orthogonal method, offers elegant synthesis solutions [67, 74]. Finally, adaptive and non-deterministic optimization techniques could be used with success [75–77].

Arrays conformed to curved platforms are, in many cases, useful in air- and space-borne vehicles. They have the characteristic of maintaining the aerodynamic integrity of the air face. Such arrays are dictated by the geometry of the supporting structure and have to meet the EM(Electromagnetic) requirements. It is noticed that much work had

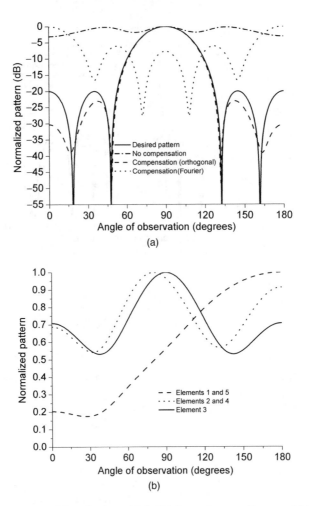

Figure 4.92 Linear array of five dipoles with 0.20λ inter-element distance. (a) Broadside Chebyshev patterns with SLL = −20 dB. (b) Patterns of the array elements

been devoted to these arrays in the past, but a general and widely accepted method for the analysis and synthesis problems is still lacking.

In most of these problems of the analysis of conformal arrays, the uniform theory of diffraction (UTD) [78, 79] or Green's function [80–86] can be used. One of the powerful techniques for the synthesis is the orthogonal method. In the next paragraphs, three different cases of conformal arrays will be presented. One involves microstrip cylindrical arrays, the other involves slot arrays on a perfectly conducting elliptic cone and the third one involves slot arrays on a perfectly conducting paraboloid.

4.11.1 Synthesis of Microstrip Cylindrical Conformal Arrays: The Orthogonal Method
Conventional tools for planar arrays are not always suitable in conformal geometries. These tools must be able to analyse the whole system, including mutual coupling, and to synthesize the array. So, it is important to find new antenna-modelling tools.

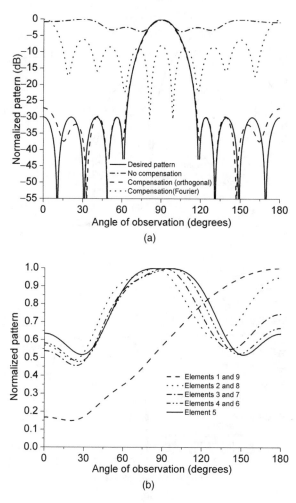

Figure 4.93 Linear array of nine dipoles with 0.20λ inter-element distance. (a) Broadside Chebyshev patterns with SLL = −20 dB. (b) Patterns of the array elements

A number of textbooks and references present extended information about conformal arrays. Several applications on aircrafts, missiles, multi-target acquisition, satellites, mobile communications, remote sensing and biomedical systems can be adopted for the above arrays. Among them, the immediate need for low profile, lightweight and simple manufacturing motivates the choice of microstrip antennas as elements of the arrays.

Of course, such a choice poses fundamental issues that must be addressed. These issues have to do with bandwidth, radiation pattern and efficiency, as well as with certain operational factors.

Microstrip antennas can be easily conformed to objects with a curved surface. Depending on the radius of curvature, different theoretical approaches can be applied. In the case of cylindrical conformal patch arrays, the cavity model and the surface current model can be used.

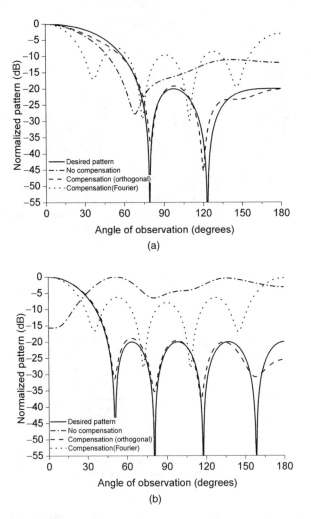

Figure 4.94   Linear array of five dipoles with inter-element distance equal to 0.20λ. (a) End-fire Chebyshev patterns (case 2) with SLL = −20 dB. (b) End-fire Chebyshev patterns (case 3) with SLL = −20 dB

In this study, the cylinder is assumed to be infinitely long and the edge diffraction from the basis to be absent. Figure 4.96 shows the position and two different probe excitations of two rectangular microstrip antennas (RMSAs).

In the analysis that follows, the two models of the RMSAs are given. On the basis of the equations of the electric field of the RMSA, the orthogonal method for the synthesis of an array composed of a set of the above antennas is presented.

Cavity Model
In the cavity model, one postulates that magnetic walls enclose the dielectric volume under the microstrip patch. In this way, a cavity is formed. The usual assumption is that because of small height, w, of the cavity, only transverse magnetic (TM) modes to a

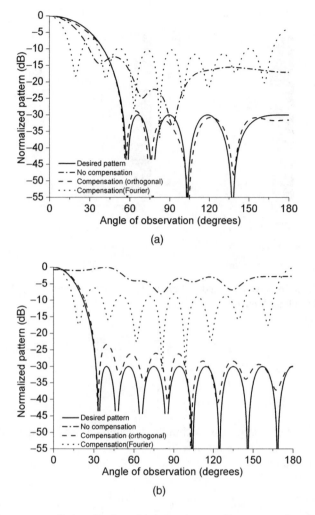

Figure 4.95 Linear array of nine dipoles with inter-element distance equal to 0.20λ. (a) End-fire Chebyshev patterns (case 2) with SLL $= -30$ dB. (b) End-fire Chebyshev patterns (case 3) with SLL $= -30$ dB

vector normal to the metallization exist. The fields in the cavity are found by performing modal expansion. The rectangular patch is assumed to consist of two axial and two circumferential slots. The external field is found from the distribution of the internal one on the side walls (slots).

The radiated far field from each mode is given by [85]

1. Axial slots:

$$E_{\phi,q}(r, \phi, \theta) = V_{pq} \frac{\exp(-j\beta r)}{2\pi^2 r R} \frac{\beta \cos\theta [\exp(j\beta z_m \cos\theta) \cos(q\pi) - 1]}{(q\pi/z_m)^2 - \beta^2 \cos^2\theta}$$

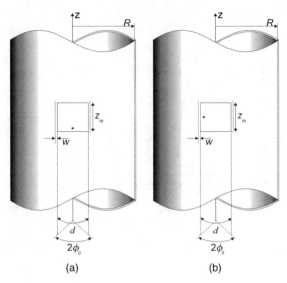

Figure 4.96 (a) Axially (vertical) and (b) circumferentially (horizontal) polarized patch on a dielectric coated circular cylinder

$$\times\, e^{-j\beta\cos\theta z_m/2} \sum_{n=0}^{\infty} \frac{\epsilon_n j^{n+1} \cos[n(\phi \pm \phi_0)]}{H_n^{(2)'}(\beta R \sin\theta)} \sin c\left(\frac{nw}{2R}\right)$$

$$\times \begin{bmatrix} 1 \\ -\cos(p\pi) \end{bmatrix} \tag{4-174}$$

where the upper and lower terms within the bracket are used for the left and right slot. $\epsilon_n = 2$ for $n > 0$ and $\epsilon_n = 1$ for $n = 0$.

2. Circumferential slots:

$$E_{\theta,p}(r,\phi,\theta) = V_{pq}\frac{\exp(-j\beta r)}{2\pi^2 r} \sin c\left(\frac{\beta w}{2}\cos\theta\right) \begin{bmatrix} -e^{-j\beta\cos\theta z_m/2} \\ \cos(\pi q)e^{j\beta\cos\theta z_m/2} \end{bmatrix}$$

$$\sum_{n=-\infty}^{\infty} \exp[jn(\phi + \phi_0)]j^n \frac{[\cos(p\pi)\exp(-jn2\phi_0) - 1]}{H_n^{(2)}(\beta R \sin\theta)\sin\theta}$$

$$\frac{n}{[\pi p/(2\phi_0)]^2 - n^2} \tag{4-175}$$

$$E_{\theta,p}(r,\phi,\theta) = \frac{V_{pq}}{\beta R}\frac{\exp(-j\beta r)}{2\pi^2 r} \sin c\left(\frac{\beta w}{2}\cos\theta\right) \begin{bmatrix} -e^{-j\beta\cos\theta z_m/2} \\ -\cos(\pi q)e^{j\beta\cos\theta z_m/2} \end{bmatrix}$$

$$\frac{\cos\theta}{\sin\theta} \sum_{n=-\infty}^{\infty} \frac{\exp[jn(\phi + \phi_0)]j^{n+1}n^2}{\sin\theta H_n^{(2)'}(\beta R \sin\theta)} \frac{[\cos(p\pi)\exp(-jn2\phi_0) - 1]}{[\pi p/(2\phi_0)]^2 - n^2}$$

$$\tag{4-176}$$

where the upper and lower terms within the bracket are used for the upper and lower slot respectively.

Surface Current Model

In the surface current model, the key assumption is that the patch metallization is replaced by a surface electric current density. Green's function, with components $G_{zz}, G_{z\phi}, G_{\phi z}$ and $G_{\phi\phi}$ as an infinite sum of Bessel functions, is used. The far field is acquired via the inverse Fourier transform by using the steepest descent method of integration. The electric field in the spectral domain is given by

$$\begin{bmatrix} E_\theta(n, k_z) \\ E_\phi(n, k_z) \end{bmatrix} = \sin(\theta)\, j^n \begin{bmatrix} -\omega\mu G_{zz}(n, k_z) & -\omega\mu G_{z\phi}(n, k_z) \\ k_0 G_{\phi z}(n, k_z) & k_0 G_{\phi\phi}(n, k_z) \end{bmatrix} \cdot \begin{bmatrix} \tilde{J}_z(n, k_z) \\ \tilde{J}_\phi(n, k_z) \end{bmatrix} \quad (4\text{-}177)$$

where $\tilde{J}_z$ and $\tilde{J}_\phi$ are the z and ϕ components in the spectral domain of the surface current density on the patch.

By using the inverse Fourier transform, we find

$$E_\theta = -\omega\mu \frac{e^{-j\beta r}}{\pi r} \sin\theta \sum_{n=-\infty}^{n=+\infty} e^{jn\cdot\phi} j^n G_{zz}(n, \beta\cos\theta)\tilde{J}_z(n, \beta\cos\theta)$$

$$\quad (4\text{-}178)$$

$$E_\phi = \beta \frac{e^{-j\beta r}}{\pi r} \sin\theta \sum_{n=-\infty}^{n=+\infty} e^{jn\cdot\phi} j^n G_{\phi\phi}(n, \beta\cos\theta)\tilde{J}_\phi(n, \beta\cos\theta)$$

According to the above equations, for the axial and the circumferential slots, Figs. 4.97 and 4.98 show the field patterns computed by the cavity model using the first five modes. It must be pointed out that, usually, only the first two are the dominant ones. The size

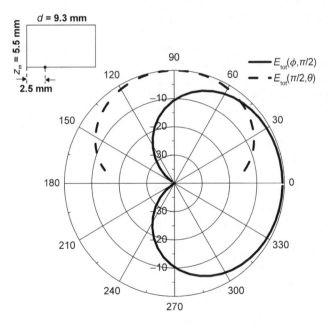

Figure 4.97 Electric field of an axially (vertical) polarized patch on a circular cylinder (five modes) for 10 GHz

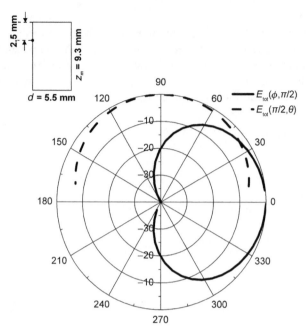

Figure 4.98 Electric field of a circumferentially (horizontal) polarized patch on a circular cylinder (five modes) for 10 GHz

of the RMSA is 9.3×5.5 mm and the feed probe is given in two different positions. The frequency of operation is 10 GHz, while layer thickness and relative dielectric constant are 1.57 and 2.2 mm respectively. The cylinder radius is 0.5 m. The modes that are used are $\{(1, 0), (0, 1), (1, 1), (2, 0), (0, 2)\}$.

The required number of terms in the field equations depends on the radius of the cylinder and the frequency of operation. The number of terms was found to be approximately $2\beta R$. This is illustrated in Fig. 4.99, where the radiation pattern of an axial microstrip antenna for various numbers of terms is shown. The MSA is the square of the 15.2 mm side on a coated cylinder with $R = 1.9\lambda$ and $f = 5.7$ GHz. The substrate thickness is $w = 1.57$ mm and $\epsilon_r = 2.32$. It is obvious that we have convergence for $n = 25 > 2\beta R \cong 24$.

Both the cavity and the surface current models are used to verify the numerical results, and the pattern is presented in Fig. 4.100. A comparison between the results of both models has shown that there is acceptable agreement.

After verifying the accuracy of the numerical procedure for the RMSA field, we proceed to the orthogonal method.

We start with the assumption that we have an N-element (RMSA) array that is conformed on a dielectric coated circular cylinder.

The electric field expression is

$$E(\theta, \phi) = \sum_{n=1}^{N} I_n [E_n^\theta(\theta, \phi)\hat{\boldsymbol{u}}^\theta + E_n^\phi(\theta, \phi)\hat{\boldsymbol{u}}^\phi] = \sum_{n=1}^{N} I_n \boldsymbol{\phi}_n(\theta, \phi) \qquad (4\text{-}179)$$

where I_n is the complex excitation and E_n^θ and E_n^ϕ are the theta- and phi-polarized electric fields of the nth (port) element of the array.

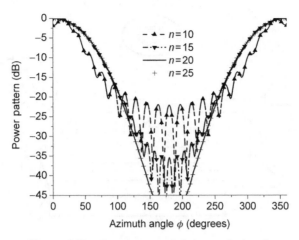

Figure 4.99   Convergence of the numerical code

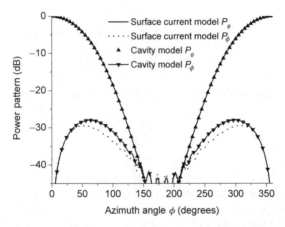

Figure 4.100   Pattern comparison by using the cavity and the surface current models

Equation (4-179) shows that the electric field is a vector of an N-dimensional vector space.

In general, basis $\{\boldsymbol{\phi}_j\}$ is not orthogonal. With a method analogous to that of the orthogonal one, we can create a new basis, where the electric field can be expressed.

With the aid of the Gram-Schmidt procedure, the orthogonal basis $\{\boldsymbol{\psi}_j\}$ is derived:

$$\boldsymbol{\psi}_j(\theta, \phi) = \sum_{i=1}^{j} C_i^{(j)} \boldsymbol{\phi}_i(\theta, \phi) \qquad (4\text{-}180)$$

In the new basis, the electric field is given by

$$\boldsymbol{E}(\theta, \phi) = \sum_{i=1}^{N} L_i \cdot \boldsymbol{\psi}_i(\theta, \phi) \qquad (4\text{-}181)$$

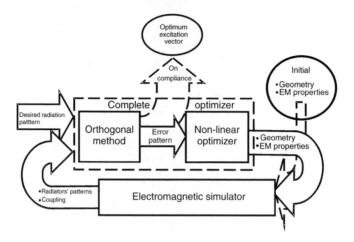

Figure 4.101   The optimization scheme for the design of an antenna array

Because of the orthogonal properties of the new basis, the amplitudes L_i are

$$L_i = <E(\theta, \phi), \boldsymbol{\psi}_i(\theta, \phi)> \tag{4-182}$$

Finally, from L_i, the desired excitations of the array elements can be found by

$$I_n = \sum_{i=n}^{N} L_i C_n^{(i)} \tag{4-183}$$

For a given antenna structure, the orthogonal method derives the excitation that minimizes the squared error between the desired and acquired field. In other words, in the excitation's solution space and in the metric defined by the LMSE (difference between acquired and desired field), the orthogonal method is a global optimizer.

A complete design procedure (optimization scheme) for an antenna array is illustrated in Fig. 4.101. In this, the system parameters are found by using an optimization scheme, which can be a deterministic (say a gradient-based one) or even a stochastic one.

Next, the design of an antenna system that consists of probe feed rectangular patches on a dielectric coated circular cylindrical conductor is given. The specifications of the system are covered in Table 4.16.

Table 4.16   Specification of an antenna array of patches positioned on a circular cylinder

Frequency		10 GHz	
Cylinder radius		.5 m	
Dielectric layer	W		ϵ_r
thickness and constant	1.57 mm		2.2
Antenna radiation pattern	θ polarization	HPBW	SLL
	Azimuth	~6°	≤ 30 dB
	Elevation	25° ~ 30°	≤30 dB

The array consists of RMSAs with the characteristics given in Fig. 4.98.

The antenna geometry can be a multi-ring conformal array.

To meet the pattern specifications, we apply the orthogonal method successively: first, for the azimuth pattern and, second, for the vertical one.

Our first concern is the θ polarized electric field.

On the horizontal plane, the electric field of a ring array with uniformly positioned RMSAs is

$$E = \sum_{n=1}^{N} I_n \boldsymbol{\phi}_n(\theta = \pi/2, \phi) = \sum_{n=1}^{N} I_n E_n^{\theta}(\theta = \pi/2, \phi) \cdot \hat{\boldsymbol{u}}_{\theta} \qquad (4\text{-}184)$$

The next step is the evaluation of the desired field. In the examples, Chebyshev polynomials are used as the desired patterns because their maximum and the SLL can be controlled. The design expression in our case is

$$T_m(x(\phi)) \approx \sum_{n=1}^{N} I_n \cdot E_n^{\theta}(\theta = \pi/2, \phi) \qquad (4\text{-}185)$$

where m is the order of the Chebyshev polynomial:

$$x(\phi) = x_0 \cdot \cos(a \cdot \sin(\phi)/2)$$

$$a = 2 \cdot \cos^{-1}\left(\frac{x_{\text{HPBW}}}{x_0}\right) / \sin\left(\frac{\text{HPBW}}{2}\right)$$

$$x_0 = \cos h\left(\frac{\cosh^{-1}(R)}{m}\right) \qquad (4\text{-}186)$$

$$x_{\text{HPBW}} = \cosh\left(\frac{\cosh^{-1}(R_{\text{HPBW}})}{m}\right)$$

$$R = 10^{SLL/20} \quad and \quad R_{\text{HPBW}} = 10^{(SLL-3)/20}$$

For $m = 7$, HPBW $\leq 6°$ and SLL ≤ -30 dB, an array with 24 patches and inter-element angle $2°$, which is equivalent to a distance of $1.1635\lambda_0$, was designed. The resulting pattern has HPBW $= 5.03°$ and SLL $= -30.7$ dB. Figure 4.102 shows the pattern, while Fig. 4.103 shows the excitation of the array.

Figure 4.103 shows that the desired SLL gives a ratio between of max to min excitation equal to 7:1. To have a smaller ratio, a larger SLL can be imposed.

A similar example of an array with a smaller number of elements ($N = 21$) and inter-element angle $= 2.5°$, which is equivalent to a distance of $1.454375\lambda_0$, is presented. The radiation pattern and the excitation of the array are given in Figs. 4.104 and 4.105.

Finally, for the vertical pattern, a similar or a simpler procedure can be applied. Figure 4.106 shows the vertical pattern in four different cases.

The orthogonal method can also be applied for the design of arrays with scanning characteristics. In this case, the maximum of the pattern will be in different directions in different instances. For a single ring array with $11(N = 11)$ RMSAs and inter-element angle $= 2.5°$, the pattern for different angles ($0°$, $20°$, $40°$, $60°$) of the direction of maximum is given in Fig. 4.107. The geometry is shown in Fig. 4.108.

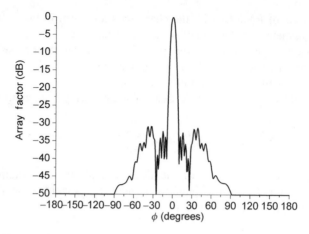

Figure 4.102   The radiation pattern of a ring array of 24 patches with SLL ≤ 30 dB

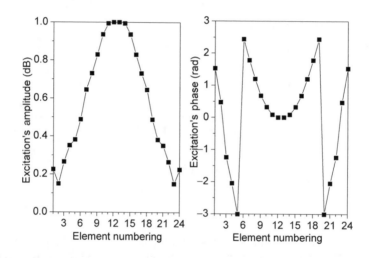

Figure 4.103   The magnitude and phase of the excitation of the array with 24 patches, with the pattern given in Fig. 4.102

From the presented patterns, it is shown that the orthogonal method can be employed to design the array. However, because of the radiation characteristics of the RMSA, the grating lobes do not give the acceptable solutions in all cases. It is obvious that up to 20° the array is acceptable. So, an approach could be to use more (say 69) RMSAs in one ring, from which only 11 are excited. In this case, a multiplexer will control the excitation. The final antenna multi-ring geometry is presented in the following figure (Fig. 4.109).

4.11.2 Synthesis of Slotted Conical Conformal Arrays: The Orthogonal Method

In this section, the orthogonal method is employed to design an array of slots on a conducting elliptic cone. The elliptic cone is an interesting geometry in antenna engineering

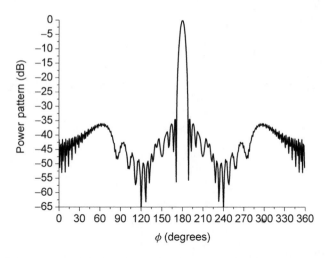

Figure 4.104   The radiation pattern of a ring array of 21 patches with SLL ≤ 30 dB

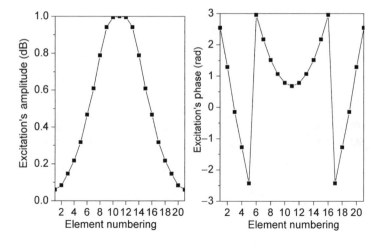

Figure 4.105 The magnitude and phase of the excitation of the array with 21 patches with the pattern given in Fig. 4.104

because it can be used in aircraft and missile applications. In the literature, one can find a series of rigorous studies related to the scattering and the slot radiation problems of a cone [80–83]. The problem of the synthesis of slot arrays conformed on a perfectly conducting elliptic cone shows an increased complexity. The same problem on a circular cone was rarely addressed in the past [87–89]. The present synthesis is based on a rigorous full-wave analysis. The Green's function of an elliptic cone can be expressed in terms of non-periodic and periodic Lamé functions. The latter are derived from the solution of the wave equation in the spheroconal coordinate system, where the elliptic cone is one of the coordinate surfaces. The ellipticity parameter $K (K = \cos(a_y)/\cos(a_x))$ and the angle

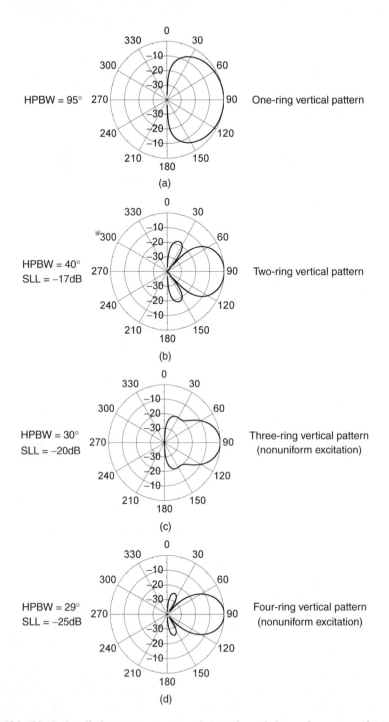

Figure 4.106 Vertical radiation patterns versus the number of rings of an array of patches. The distance between the rings is $\sim 0.6\lambda_0$

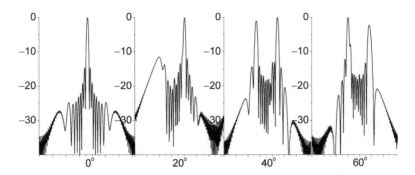

Figure 4.107 Pattern of an array for different steering angles

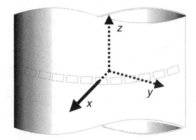

Figure 4.108 The geometry of a ring array of patches

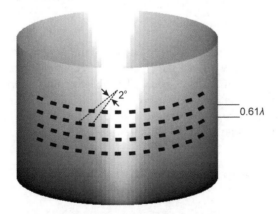

Figure 4.109 The final geometry of the conformal multi-ring array of patches

$\theta_c (\theta_c = \pi - a_x)$ are used to define the cone, and a_x, a_y are the half cone angles on the xz and yz planes respectively. Figure 4.110 shows the geometry of the cone.

The electric field of an infinitesimal magnetic dipole situated on a perfectly conducting elliptic cone can be found with the help of the $\overset{=em}{\Gamma} (R, R')$ Green's function [83], which

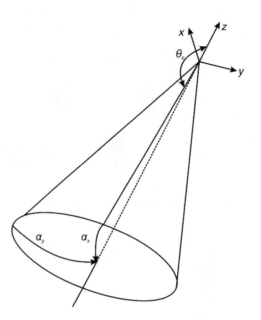

Figure 4.110   Geometry of an elliptic cone

is given by

$$\overline{\overline{\Gamma}}^{em}(\boldsymbol{R},\boldsymbol{R}') = j\beta^2 \cdot \sum_q \left[\frac{N_{q_2}^{\mid}(\boldsymbol{R}')M_{q_2}^{\parallel}(\boldsymbol{R}')}{\Lambda_{q_2}} + \frac{M_{q_1}^{\mid}(\boldsymbol{R}')N_{q_1}^{\parallel}(\boldsymbol{R})}{\Lambda_{q_1}} \right] \tag{4-187}$$

$\boldsymbol{R}'$ and $\boldsymbol{R}$ are the position vectors of the source and the field point, respectively, $\boldsymbol{M}$ and $\boldsymbol{N}$ are the wave functions [79], q_i are the eigenvalues, Λ_{q_i} are the normalization constants and β is the wave number of the free space. The subscript 1 or 2 corresponds to Dirichlet or Neumann boundary conditions, respectively, while the superscripts $\mid$ and $\parallel$ represent a standing and an outward propagating wave respectively. The electric field radiated from a slot positioned on the cone is derived from the following equation.

$$\boldsymbol{E}(\boldsymbol{R}) = \int_S \boldsymbol{E}(\boldsymbol{R}') \times \hat{\boldsymbol{n}} \cdot \overline{\overline{\Gamma}}^{em}(\boldsymbol{R},\boldsymbol{R}') \cdot ds \tag{4-188}$$

where $\boldsymbol{E}(\boldsymbol{R}') \times \hat{\boldsymbol{n}}$ is the magnetic current density of the slot and S is its cross section.

The mutual coupling between two infinitesimal magnetic dipoles situated on a conducting cone is found by using the $\overline{\overline{\Gamma}}^{hm}(\boldsymbol{R},\boldsymbol{R}')$ Green's function, which is given by

$$\overline{\overline{\Gamma}}^{hm}(\boldsymbol{R},\boldsymbol{R}') = -\beta\omega\epsilon \cdot \sum_q \left[\frac{M_{q_2}^{\mid}(\boldsymbol{R}')M_{q_2}^{\parallel}(\boldsymbol{R})}{\Lambda_{q_2}} + \frac{N_{q_1}^{\mid}(\boldsymbol{R}')N_{q_1}^{\parallel}(\boldsymbol{R})}{\Lambda_{q_1}} \right] \tag{4-189}$$

The magnetic field radiated from a slot is

$$H(R) = \int_S E(R') \times \hat{n} \cdot \overset{=hm}{\Gamma}(R,R') \cdot ds \qquad (4\text{-}190)$$

It is noticed that the computation of the eigenvalues and the corresponding eigenfunctions is a difficult task. Each eigenfunction is expressed as the product of a spherical Bessel function and a Lamé product. The Lamé product is expressed as a product of a periodic and a non-periodic Lamé function and may fall into four functional types. The difficulty in deriving the numerical solution does not reside in the complexity of the formulation but rather in the computation of the functions employed in the non-periodic term.

A cone with an ellipticity parameter $K^2 = 0.6$ and $\theta_c = 160°$ has been used in the design examples. In this cone, a typical diagram of the Neumann and Dirichlet eigenvalues is given in Fig. 4.111. λ and v are the eigenvalues of the coupled Lamé differential equations with periodic and non-periodic solutions. The parameter m indicates the ordered numbering of the $\lambda(v)$ eigenvalue curves corresponding to the periodic Lamé equation.

The eigenfunctions help derive the dyadic Green's functions through a proper series summation. The necessary number of terms in the summation depends on the position R' of the slot and extends, in the radiation case, to the eigenvalue v, for which $v \geq \beta \cdot R'$.

Implementation of the presented theory leads to the analysis of slot arrays positioned on the cone.

One slot is positioned at $r = 2\lambda$ from the tip of the cone, while the other moves away from the tip by a distance of $2.5\lambda–8\lambda$. The centres of both slots reside on the xz plane (see Fig. 4.110). The slots are assumed to be rectangular with dimensions $\lambda/2 \times \lambda/5$. The admittances are calculated by using [79]

$$Y_{ji} = \frac{\underset{Si\ Sj}{\iint\iint} d\boldsymbol{p}_m(\boldsymbol{r}_j) \cdot \overset{=hm}{\Gamma}(\boldsymbol{r}_j,\boldsymbol{r}_i) \cdot d\boldsymbol{p}_m(\boldsymbol{r}_i)}{V_i \cdot V_j} \qquad (4\text{-}191)$$

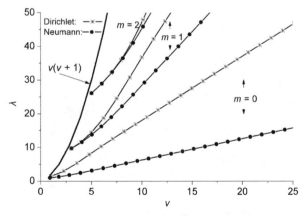

Figure 4.111 Eigenvalues of the coupled Lamé equations for $K^2 = 0.6$ and $\theta_c = 160°$

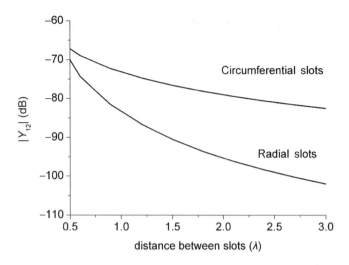

Figure 4.112   The mutual admittance between two slots on an elliptic cone with $K^2 = 0.6$ and $\theta_c = 160°$ in the circumferential and radial case

In Equation (4-191), $S_i = a_i b_i$ and $S_j = a_j b_j$ are the areas and a_i, b_i, a_j, b_j are the dimensions of the two slots. $d\boldsymbol{p}_m$ is the infinitesimal vector component of the dominant mode's magnetic current density (dominant mode approximation). The normalization constants obey the $V_{i,j} = \sqrt{a_{i,j} b_{i,j}/2}$.

In Fig. 4.112, both the mutual admittance of the circumferential and the radial slot cases are presented. The coupling level in the circumferential case is higher than the one in the radial case.

Figure 4.113 presents the radiation pattern on the xz plane for a circumferential (Fig. 4.113a, E-plane) and a radial (Fig. 4.113b, H-plane) slot positioned at $r' = \lambda$ (dotted line) and $r' = 10\lambda$ (straight line) away from the tip. Both slots exhibit similar behaviour. However, in the case of a circumferential slot at $r' = \lambda$, the omnidirectional behaviour of the slot is changed because of the tip diffraction. As the distance of the slot from

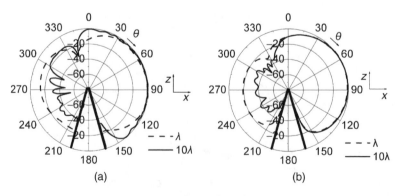

Figure 4.113   The radiation pattern of (a) a circumferential and (b) a radial slot, positioned at $r' = \lambda$ (dotted line) and $r' = 10\lambda$ (straight line) on an elliptic cone with $K^2 = 0.6$ and $\theta_c = 160°$

the tip increases, the pattern for the circumferential slot gradually becomes almost omnidirectional in the lit region. The pattern of a radial slot obeys a cosine function in the same region, even at a small distance from the tip. This is because there is no incident field from the slot to the tip. The above results are qualitatively similar to the ones in [90] for a circular cone, where the existence of an omnidirectional pattern and a cosine pattern in the lit region is observed.

The far electric field of a conformal array of slots can be expressed in matrix form as follows.

$$E(\theta, \phi) = [F]^T \cdot [W] \qquad (4\text{-}192)$$

The nth element, $F_n(\theta, \phi)$, of the vector $[F]$ is the far electric field produced by the nth current mode. The nth weight, W_n, in $[W]$ is the respective complex amplitude-excitation. We adopt the inner product:

$$\langle F_n | F_m \rangle = \int_0^{2\pi} \int_0^{\pi} F_n(\theta, \phi) \cdot F_m^*(\theta, \phi) \cdot w(\theta, \phi) \cdot \sin\theta \cdot d\theta \cdot d\phi = K_{nm} \qquad (4\text{-}193)$$

where w is the appropriate weight function.

By employing the Gram-Schmidt procedure, the orthogonal method creates an orthogonal basis. By using the notation of a lower triangular matrix $[C]$ we have

$$\{F_n(\theta, \phi)\} \xrightarrow{[c]} \{\hat{F}_n(\theta, \phi)\} \Leftrightarrow$$

$$\hat{F}_n(\theta, \phi) = \sum_{v=1}^{n} C_v^{(n)} \cdot F_v(\theta, \phi) \Leftrightarrow [\hat{F}] = [C] \cdot [F] \qquad (4\text{-}194)$$

The total field over the new basis is

$$E(\theta, \phi) = \sum_{n=1}^{N} \hat{\mathbf{F}}_n(\theta, \phi) \cdot B_n \Leftrightarrow E(\theta, \phi) = [\hat{\mathbf{F}}]^T \cdot [B] \qquad (4\text{-}195)$$

From the desired field $E^{\text{desired}}(\theta, \phi)$, the B_n coefficients are found to be

$$B_n = \langle \hat{\mathbf{F}}_n | E^{\text{desired}} \rangle \Leftrightarrow [B] = \langle [\hat{\mathbf{F}}] | E^{\text{desired}} \rangle \qquad (4\text{-}196)$$

From $[B]$ the desired excitations $[W]$ can be easily calculated and is given by

$$[B] \xrightarrow{[c]} [W] \Leftrightarrow$$

$$W_n = \sum_{j=n}^{N} C_n^{(j)} \cdot B_j \Leftrightarrow [W] = [C]^T \cdot [B] \qquad (4\text{-}197)$$

In the above derivation, no coupling was taken into account. The inclusion of the coupling parameters depends on how Equation (4-192) is constructed. For the design examples that follow, each of the N slots that comprise the array is fed by its own waveguide. We assume that an incident to the slot and a reflected wave mode exist in each waveguide (dominant mode approximation). The array with the feeding network can be modelled as a multiport network with N ports. If we assume matching conditions where there are no reflected wave modes, the far field of the array is

$$E(\theta, \phi) = \sum_{n=1}^{N} F_n(\theta, \phi) \cdot E_n^+ = [F]^T[E^+] \tag{4-198}$$

$F_n(\theta, \phi)$ is the far field of the nth slot and E_n^+ is the excitation amplitude of the dominant mode incident to this slot from its feeding waveguide. $[F]$ and $[E^+]$ are the respective column vectors with N elements.

If mutual coupling is taken into account, Equation (4-198) is modified to

$$E(\theta, \phi) = \sum_{n=1}^{N} \tilde{F}_n(\theta, \phi) \cdot \tilde{E}_n^+ = [\tilde{F}]^T[\tilde{E}^+] \tag{4-199}$$

where $\tilde{E}_n^+$ is the excitation amplitude of the dominant mode, but $\tilde{F}_n(\theta, \phi)$ is the electric field of the array when only the mth slot is excited. The latter is given by

$$\tilde{F}_m(\theta, \phi) = \sum_{n=1}^{N} F_n(\theta, \phi)(S_{nm} + I_{nm}) \Leftrightarrow [\tilde{F}]^T = [F]^T([S] + [I]) \tag{4-200}$$

where S_{nm} and I_{nm} are the nm elements of the scattering matrix $[S]$, and of the identity matrix $[I]$, respectively [91]. The scattering matrix obeys the equation [68]

$$[S] = ([Y_{\text{guide}}] + [Y])^{-1}([Y_{\text{guide}}] - [Y]) \tag{4-201}$$

In Equation (4-201), $[Y]$ is the matrix of the mutual admittances of the slots, while $[Y_{\text{guide}}]$ is a diagonal matrix of the intrinsic admittances of the feeding waveguides.

By implementing the orthogonal method to Equations (4-198) and (4-199), the $[E^+]$ and $[\tilde{E}^+]$ are obtained. It is interesting to note that the relation between them is

$$[\tilde{E}^+] = ([S] + [I])^{-1}[E^+] \tag{4-202}$$

$([S] + [I])^{-1}$ contains the coupling compensation factors. These factors can be trivially generalized even if modes more than the dominant ones are allowed to exist in the feeding waveguides.

In the examples that follow, the slots are considered to be rectangular with dimensions $\lambda/2 \times \lambda/5$. The field in each slot can be well approximated with a simple cosine distribution, that is, the so-called one-mode approximation. Each slot is considered to be connected to an RF chain consisting of a precision attenuator and a precision phase shifter or, alternatively, of a vector modulator. It is assumed that coupling between the slots comes from the space where they radiate.

Two different arrays will be presented. One is composed of circumferential (Fig. 4.114a) and the other of quasi-radial slots (Fig. 4.115a). In the case of uniform excitation, the resulting far-field patterns are given in Figs. 4.114b and 4.115b respectively. In both cases, the broadside character with an SLL ≈ -13.5 dB of the arrays is observed. As expected, in the radial case, the HPBW is smaller than that of the circumferential one. For an array with strict requirements (low SLL, small HPBW and scanning capability), non-uniform excitation is needed, and this will be provided by the orthogonal method.

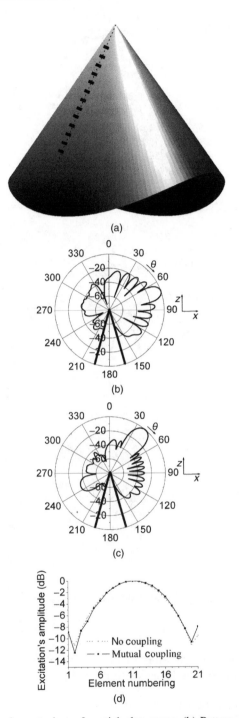

Figure 4.114   (a) A 21-element circumferential slot array. (b) Pattern of a uniform excitation. (c) Pattern of a Chebyshev array. (d) Chebyshev excitation no coupling (N.C.) and mutual coupling (M.C.) between slots

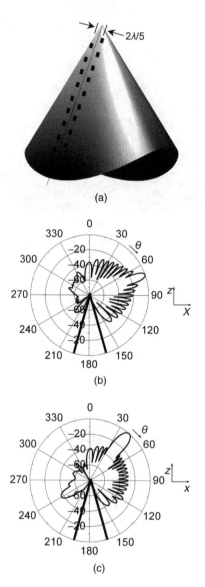

Figure 4.115 (a) A 20-element quasi-radial slot array. (b) Pattern of a uniform excitation. (c) Pattern of a Chebyshev array

Let us suppose that we have an array of 21 circumferential slots positioned on a cone with $K^2 = 0.6$ and $\theta_c = 160°$ at distances from the tip:

$$r_n = (2 - 0.3 + n \cdot 0.3) \cdot \lambda, \quad n = 1, \dots, 21 \qquad (4\text{-}203)$$

The position r_1 of the centre of the first slot is 2λ, while the last one is located at 8λ away from the tip. The radiation pattern of a T_{20} Chebyshev array (Fig. 4.114c) with

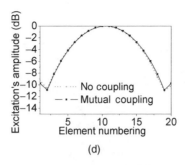

(d)

Figure 4.115 (d) Chebyshev excitation no coupling (N.C.) and mutual coupling (M.C.) between slots

SLL of −30 dB is presented. The HPBW is approximately equal to 12°, while the desired main lobe direction is at 40° away from the z-axis (30° far from the broadside direction).

Figure 4.115a shows an array consisting of 20 quasi-radial slots positioned on a $K^2 = 0.6$ and $\theta_c = 160°$ elliptic cone at distances

$$r_n = (2 - 0.5 + n \cdot 0.5) \cdot \lambda, \quad n = 1, \ldots, 20 \tag{4-204}$$

The first slot is 2λ, while the last one is 11.5λ away from the tip. The pattern of a T_{19} Chebyshev (Fig. 4.115c) array with SLL of −30 dB is given. The HPBW is approximately equal to 8°, while the main lobe direction is at 40° far from the z-axis (30° far from the broadside direction).

During the design of the above slot arrays, it was found that the main lobe can be directed up to ±30° from the broadside direction by taking the suitable phase differences. It is also important to note that the design can be based on the desired response in the lit region ('design-on-the-lit-region' concept) because the field in the shadow region exhibits extremely low power level.

Figures 4.114d and 4.115d present the excitation amplitudes of the circumferential and the radial slot Chebyshev arrays. The minimum relative excitation values are $>−12.5$ dB, which classify the solution as feasible.

4.11.3 Synthesis of Slot Arrays on a Perfectly Conducting Paraboloid: The Orthogonal Method

In the present section, the UTD for the analysis of slot arrays and the orthogonal method for their synthesis are employed. Slot arrays positioned on a perfectly conducting paraboloid will be presented.

Mutual coupling between two slots positioned on the paraboloid is derived and the UTD-OM analysis-synthesis technique is introduced for the design of an array.

It is well known that UTD is an asymptotic technique that can be easily implemented with relatively small requirements on computational recourses. However, UTD is problematic when it is treated in caustics and paraxial regions [78, 79]. In the case of a paraboloid (Fig. 4.116), three detachment points, connecting a point source to the far field, must be included in the calculation.

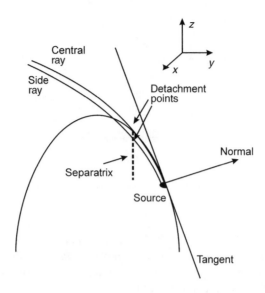

Figure 4.116   Ray paths on a paraboloid surface

The geometry under study is a conducting paraboloid of revolution, which is a coordinate surface of the parabolic coordinate system. The parametric equations of the above system are [92]

$$x = au \cos v \quad y = au \sin v \quad z = -u^2 \tag{4-205}$$

where a is the sharpness/flatness parameter and u and v are the independent variables of the surface. The receiver is considered to be in the direction $\hat{\mathbf{d}}$ towards the infinity on the zy plane. Its position is defined by the angle θ, which is $(\theta = \cos^{-1}(\hat{\mathbf{z}}, \hat{\mathbf{d}}))$.

One of the main steps in the process is the determination of the geodesic paths on the parabolic surface. The key step in computing the geodesics is the determination of the shadow and light separatrix. In fact, the separatrix is the locus of the shadow boundary. If the positions of the source and the receiver are given, the separatrix contains all possible detachment points (Fig. 4.116). A ray follows a geodesic path on the surface, connecting the source with the detachment point. It is then transmitted to the receiver along a straight line that coincides with the direction of the tangent to the surface at the detachment point. For the present geometry, it is proven that more than one detachment point fulfils the above condition, meaning that more than one ray reaches the receiver. An example that clarifies this ray-tracing mechanism follows. In Fig. 4.117, the ray paths leaving the source positioned at $(u, v) = (2, -\pi/2)$ are presented. The sharpness/flatness parameter is set at unity.

In the first case, these paths end at the separatrix defined for a receiver angle at $\theta = 20°$. In the second case, the paths end at the separatrix defined for $\theta = 50°$. From the rays reaching the separatrix, only one ray in general, and in some cases up to three, will be directed to the receiver. In fact, one ray reaches the receiver in the first case (receiver at $\theta = 20°$) and three in the second case (receiver at $\theta = 50°$). The ray that travels on the high path (central ray) always reaches the receiver, while, in the second case, two identical side rays also contribute to the field. It is obvious that there must be an angle

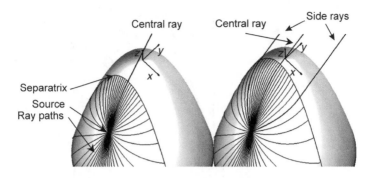

Figure 4.117   Geodesics and separatricies on the paraboloid – cases 1 and 2

between 20 and 50 degrees for the geometry under study at which one ray may break into three and a caustic is generated.

In the present case, the concept of equivalent currents is adopted [93]. Correction of caustics is made in order to study radiation from slots positioned on convex surfaces. In the shadow region of the source, the far field is due to creeping wave contribution. The expression of this contribution is given in the form of a line integral over the separatrix. In other words, the evaluation of the far field (in the shadow region) produced by a slot on a convex surface can be regarded as a two-step process. Initially, the source produces equivalent magnetic currents on the separatrix through the mechanism of the creeping waves. Then, the separatrix, as a curved-line magnetic source, radiates the far field at the receiver.

Because of equivalent magnetic currents $M(\ell)$ over the separatrix, the electric field can be given by [93]

$$E = -\frac{j\beta}{4\pi} \int_c (\hat{\mathbf{d}} \times M(\ell)) \cdot \frac{e^{-jkR(\ell)}}{R(\ell)} \cdot d\ell \qquad (4\text{-}206)$$

where $\hat{\mathbf{d}}$ is the unit vector towards the receiver, ℓ is the abscissa (the arc length) and $d\ell$ is the infinitesimal length on the separatrix. R is the distance between a point on the separatrix and the receiver.

The key step in the evaluation process is the determination of the equivalent magnetic currents, which are

$$M(\ell) = (\hat{\mathbf{d}} \times E^d(\ell))/\sqrt{\frac{-j\beta}{8\pi}} \qquad (4\text{-}207)$$

where $E^d(\ell)$ is the field that, having been produced by the source and having travelled in the form of a creeping wave, leaves the paraboloid surface detached from a point of the corresponding separatrix.

The parameter of the paraboloid is set at unity for the radiation pattern calculations hereafter.

The geometry and the θ polarized far-field pattern on the zy plane (E-plane) of a circumferential slot are illustrated in Figs. 4.118 and 4.119 respectively. In Fig. 4.120, the ϕ polarized far-field pattern on the zy plane (H-plane) of an axial slot is shown. In

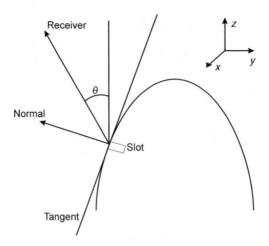

Figure 4.118   Single circumferential slot radiator on the paraboloid

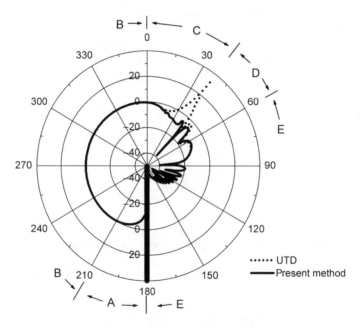

Figure 4.119   The θ polarized far-field pattern of an axial slot on the conducting paraboloid

both cases, the slot dimensions are $\lambda/2 \times \lambda/5$, and the centre of each one is located at

$$u = 2 \quad v = -\pi/2 \tag{4-208}$$

The dotted line shows the UTD results, while the continuous line illustrates the equivalent current results. Away from the caustic, the results coincide. It is obvious that an equivalent current approach does not show any divergence in the caustic region. This is important for the diffraction problem in curved surfaces.

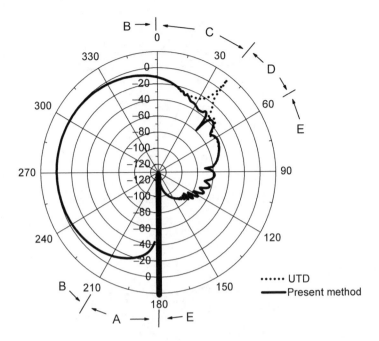

Figure 4.120   The ϕ polarized far-field pattern of an axial slot on the conducting paraboloid

The entire pattern can be partitioned into five regions, namely

A. Shadow region – one ray
B. Lit region – one ray
C. Shadow region – one ray – shadow region of the caustic
D. Shadow region – three rays – lit region of the caustic
E. Shadow region – three rays

In the following text, results of the mutual coupling between two slots positioned on a perfectly conducting paraboloid is presented. The slot dimensions are $\lambda/2 \times \lambda/5$ and the parameter a is set at 5. We consider the position of the first slot to be fixed at ($u_1 = 2$, $v_1 = -\pi/2$) and the second slot to be moving on the path $u_2 \in (1, 3)$ and $v_2 = v_1 + \phi$. (See Fig. 4.121).

We use the azimuth angle, ϕ, as a parameter. The computed results are presented in Fig. 4.122. It is assumed that each slot is fed by a rectangular waveguide in the dominant vector mode.

The far electric field of an array of N slots can be expressed by

$$E(\theta, \phi) = \sum_{n=1}^{N} F_n(\theta, \phi) \cdot V_n \tag{4-209}$$

where $F_n(\theta, \phi)$ is the far electric field and V_n is the complex amplitude of the nth current mode. The magnetic current modes are used to model the slot radiators. Equation (4-209)

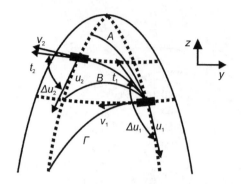

Figure 4.121   Two slots on a conducting paraboloid

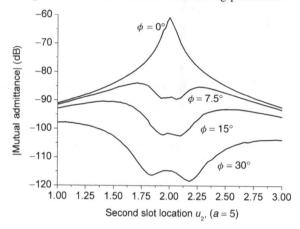

Figure 4.122   Mutual coupling between the slots of Fig. 4.121 where $u_1 = 2$, $v_1 = -\pi/2$ and $v_2 = v_1 + \phi$

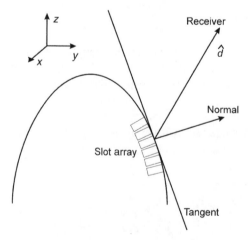

Figure 4.123   The geometry of a slot array with circumferential slots

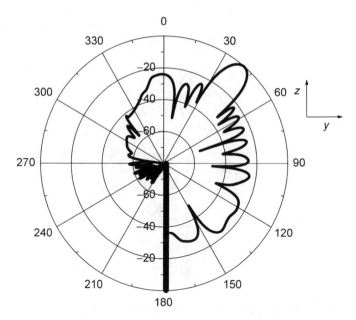

Figure 4.124   Chebyshev pattern with maximum at $\theta = 40°$

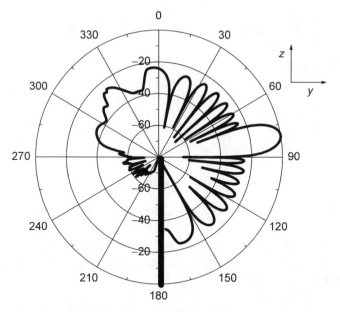

Figure 4.125   Chebyshev pattern with maximum at $\theta = 80°$

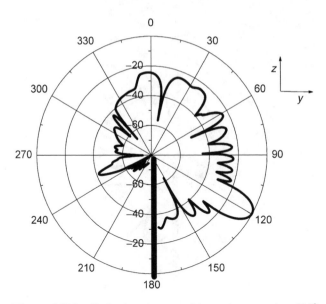

Figure 4.126   Chebyshev pattern with maximum at $\theta = 120°$

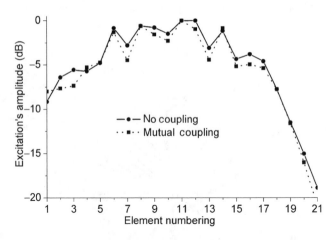

Figure 4.127   Excitation of slots with no coupling (N.C.) and with mutual coupling (M.C)

shows that the far field is expressed on the basis $\boldsymbol{F}_n(\theta, \phi)$. Following the same procedure as in the previous section, we can apply the orthogonal method to derive the complex amplitudes of the current modes.

The slot array under study is given in Fig. 4.123.

The array consists of 21 circumferential slots. The slots are rectangular with narrow width $\lambda/5$ and length equal to $\lambda/2$. The field in each slot can be sufficiently approximated with a simple cosine distribution, that is, the so-called one-mode approximation.

An array with strict requirements (low SLL, small HPBW and scanning capability) needs non-uniform excitation. For an array with SLL $= -25$ dB, a Chebyshev pattern of the 20th order is chosen.

The radiation patterns with different directions of maximum are presented in Figs. 4.124–4.126.

The main-beam direction of the above patterns is obtained by proper complex slot excitation coefficients. For example, to obtain a maximum radiation at $\theta = 40°$, we have excitations that are given in Fig. 4.127.

Various arrays can be designed as an application of the UTD-OM technique. The above method is a useful tool to synthesize arrays conformal to convex surfaces with strict radiation pattern requirements.

References

[1] I. Wolff, "Determination of the Radiating System Which Will Produce a Specified Directive Characteristic," *Proc. IRE*, Vol. **25**, pp. 630–643, 1937.

[2] S.A. Schelkunoff, "A Mathematical Theory of Linear Arrays," *Bell Syst. Tech. J.*, Vol. **22**, pp. 80–107, January 1943.

[3] C.L. Dolph, "A Current Distribution for Broadside Arrays Which Optimizes the Relationship Between Beamwidth and Side-Lobe Level," *Proc. IRE*, Vol. **34**, pp. 335–348, June 1946.

[4] H.J. Riblet, "Discussion on a Current Distribution for Broadside Arrays Which Optimizes the Relationship between Beam Width and Side-Lobe Level,*Proc. IRE*, Vol. **35**, pp. 489–492, 1947.

[5] P.M. Woodward, "A Method of Calculating the Field over a Plane Aperture Required to Produce a Given Polar Diagram," *J.I.E.E*, Vol. **93**, Part IIIa, pp. 1554–1558, 1946.

[6] P.M. Woodward and J.D. Lawson, "The Theoretical Precision with Which an Arbitrary Radiation Pattern May Be Obtained from a Surface Finite Size," *J.I.E.E.*, Vol. **95**, Part III, pp. 363–370, 1948.

[7] G.J. Van der Maas, "A Simplified Calculation for Dolph-Tchebysheff Arrays," *J. Appl. Phys.*, Vol. **25**, pp. 121–124, January 1954.

[8] T.T. Taylor, "Dolph Arrays of Many Elements," Tech. Memo, No. 320, Hughes Aircraft Company, August 18, 1953.

[9] T.T. Taylor, "Design of Line Source Antennas for Narrow Beamwidth and Low Side Lobe Level," *Trans. IRE*, Vol. **AP-3**, pp. 16–28, January 1955.

[10] D.K. Cheng and M.T. Ma, "A New Mathematical Approach in Linear Array Analysis," *Trans. IRE*, Vol. **AP-8**, No. 3, pp. 225–259, May 1960.

[11] H. Unz, *Linear Arrays with Arbitrarily Distributed Elements*, University of California, Berkeley, Electronic Research Lab., Report Series No. 60, Issue No. 168, November 2, 1956.

[12] H. Unz, *Multi-Dimensional Lattice Arrays with Arbitrarily Distributed Elements*, University of California, Berkeley, Electronic Research Lab., Rep. Series No. 60, Issue No. 172, December 19, 1956.

[13] H. Unz, "Linear Arrays with Arbitrarily Distributed Elements," *Trans. IRE*, Vol. **AP-8**, pp. 222–223, March 1960.

[14] D.D. King, R.F. Packard and R.K. Thomas, "Unequally Spaced, Broadband Antenna Arrays," *Trans. IRE*, Vol. **AP-8**, pp. 380–384, July 1960.

[15] R.H. DuHamel and G.G. Chadwick, *Frequency Independent Antennas*, eds. R. Johnson and H. Jasik, Chapter 14 in *Antenna Engineering Handbook*, McGraw-Hill Book Company, New York, 1992.

[16] R.F. Harrington, "Side Lobe Reduction of Nonuniform Element Spacing," *Trans. IRE*, Vol. **AP-9**, pp. 187–192, March 1961.

[17] H. Unz, "Nonuniform Arrays with Spacings Larger than One Wavelength," *Trans. IRE*, Vol. **AP-10**, pp. 647–648, September 1962.

[18] J.D. Bruce and H. Unz, "Broadband Nonuniformly Spaced Arrays," *Trans. IRE*, Vol. **AP-10**, p. 228, February 1962.

[19] U.L. Pokrovskii, "General Method of Seeking the Optimum Distribution of Linear Antennas," *Doklady AN USSR*, Vol. **138**, No. 3, pp. 584–586, 1961.

[20] F.W. Brown, "Note on Nonuniformly Spaced Arrays," *Trans. IRE*, Vol. **AP-10**, pp. 639–640, September 1962.

[21] H. Unz. *New Methods for Synthesis of Nonuniformly Spaced Antenna Arrays*, Antenna Lab., Report 1423-1, Ohio State University, Columbus, Ohio, pp. 1–31, November 15, 1962.

[22] Y.T. Lo, "A Spacing Weighted Antenna Array," *Trans. IRE*, Vol. **AP-10**, No. 3, p. 353, March, 1962.

[23] H. Unz, "Nonuniformly Spaced Arrays: The Orthogonal Method," *Proc. IEEE*, Vol. **54**, No. 1, pp. 53–54, January 1966.
[24] A.I. Uzkov, "An Approach to the Problem of Optimum Directive Antenna Design," *Doklady AN USSR*, Vol. **53**, No. 1, pp. 35–38, 1946.
[25] J.N. Sahalos, "The Orthogonal Method of Nonuniformly Spaced Arrays," *Proc. IEEE*, Vol. **62**, No 2, pp. 281–282, 1974.
[26] J.N. Sahalos, "A Solution of the General Nonuniformly Spaced Antenna Array," *Proc. IEEE*, Vol. **22**, pp. 1292–1294, 1974.
[27] J.N. Sahalos, K. Melidis and S. Lampou, "On the Optimum Directivity of General Nonuniformly Spaced Broadside Arrays of Dipoles," *Proc. IEEE*, Vol. **22**, pp. 1706–1708, 1974.
[28] J.N. Sahalos, "On the Design of Antennas Consisting of Arbitrary Oriented Dipoles," *IEEE Trans. Antennas Propag.*, Vol. **24**, pp. 106–109, 1976.
[29] J.N. Sahalos, "A Solution of Nonuniformly Linear Array with the Help of the Chebyshev Polynomials," *IEEE Trans. Antennas Propag.*, Vol. **24**, pp. 239–242, 1976.
[30] J.N. Sahalos and H. Papadimitraki-Chlichlia, "On the Synthesis of Nonuniform Planar Antenna Arrays," *Sci. Annals Fac. Phys. Mathem., Univ. Thessaloniki*, Vol. **16**, pp. 138–145, 1976.
[31] J.N. Sahalos, K. Melidis and H. Papadimitraki-Chlichlia, "On the General Antenna Array Synthesis," *Sci. Annals Fac. Phys. Mathem., Univ. Thessaloniki*, Vol. **16**, pp. 203–217, 1976.
[32] J.N. Sahalos, "On the Optimum Directivity of Antennas Consisting of Arbitrary Oriented Dipoles," *IEEE Trans. Antennas Propag.*, Vol. **24**, pp. 322–327, 1976.
[33] J.N. Sahalos and H. Papadimitraki-Chlichlia, "A Solution of Radiation Pattern Synthesis of Arbitrary Antennas," *Sci. Annals Fac. Phys. Mathem., Univ. Thessaloniki*, Vol. **16**, pp. 345–353, 1976.
[34] J.N. Sahalos, "Synthesis and Optimization for Arrays of Nonparallel Wire Antennas by the Orthogonal Method," *IEEE Trans. Antennas Propag.*, Vol. **26**, pp. 886–891, 1978.
[35] K. Siakavara and J.N. Sahalos, "A Hybrid MM-GTD Technique for Optimization of the Power Gain of Arrays of Wire Antennas Near an Elliptic Cylinder," *Proc. IEEE*, Vol. **73**, pp. 1426–1428, 1985.
[36] K. Siakavara and J.N. Sahalos, "A Simplification of the Synthesis of Parallel Wire Antenna Arrays," *IEEE Trans. Antennas Propag.*, Vol. **37**, pp. 936–940, 1989.
[37] T. Samaras and J.N. Sahalos, "On the Synthesis of Hyperthermia Arrays by the Orthogonal Method," *Innovation et Technologie en Biologie et Medecine*, Vol. **17**, pp. 281–291, 1996.
[38] G. Miaris and J.N. Sahalos, "The Orthogonal Method for the Geometry Synthesis of a Linear Antenna Array," *IEEE Antennas Propag. Mag.*, Vol. **41**, pp. 96–99, 1999.
[39] C. Tsironas, T.N. Kaifas and J.N. Sahalos, "On the Design of Conformal Microstrip Antenna Arrays by the Orthogonal Method," *Proceedings of the 7th International Conference on Advances in Communications and Control (COMCON 7) - Telecommunications/Signal Processing*, Rethymnos, Greece p. 205, 1999.
[40] G.S. Miaris, H. Margaritis, S. Goudos and J.N. Sahalos, "OrthoSynthesis: On the Beamforming from a Set of Uniform Linear Arrays by the Orthogonal Method," *AP 2000 Millennium Conference on Antennas and Propagation*, Davos (Switzerland), 2000.
[41] T.N. Kaifas, C. Koukourlis and J.N. Sahalos, "On the Beam Steering of Cylindrical Conformal Microstrip Antenna Arrays," *AP 2000 Millennium Conference on Antennas and Propagation*, Davos (Switzerland), 2000.
[42] T.N. Kaifas and J.N. Sahalos, "On the Design of Cylindrical Conformal Microstrip Antenna Arrays," *Proceedings of the 2nd International Symposium of BSUAE*, Xanthi (Greece), 2000.
[43] G.S. Miaris and J.N. Sahalos, "On the Independent Amplitude Level Reduction of a Broadside Linear Array by the Orthogonal Method," *Proceedings of the 2nd International Symposium of BSUAE*, Xanthi (Greece), 2000.
[44] S.K. Goudos, G.S. Miaris and J.N. Sahalos, "On the Quantized Excitation and the Geometry Synthesis of a Linear Array by the Orthogonal Method," *IEEE Trans. Antennas Propag.*, Vol. **49**, pp. 298–303, 2001.
[45] G.S. Miaris, S.K. Goudos, Chr. Iakovidis, E. Vafiadis and J.N. Sahalos, "ORthogonal Advanced Methods for Antennas – The ORAMA Computer Tool," *IEEE Antennas Propag. Mag.*, Vol. **44**, pp. 62–74, 2002.
[46] H. Margaritis, S.K. Goudos, G.S. Miaris and J.N. Sahalos, "Orthosynthesis: On the Beamforming from a Set of Uniform Linear Arrays by the Orthogonal Method," *XXVIIth General Assembly of the International Union of Radio Science, paper 767 (electronic format)*, Maastricht (Netherlands), 2002.
[47] Z.D. Zaharis, E. Vafiadis and J.N. Sahalos, "Antenna Array Synthesis by Combining the Measured Element Patterns and the Orthogonal Method," *XXVIIth General Assembly of the International Union of Radio Science, paper 1006 (electronic format)*, Maastricht (Netherlands), 2002.

[48] T.N. Kaifas, T. Samaras, E. Vafiadis and J.N. Sahalos, "A UTD-OM Technique to Design Slot Arrays on a Perfectly Conducting Paraboloid," *Proceedings of the 3rd European Workshop on Conformal Antennas*, pp. 29–32, Bonn (Germany), 2003.

[49] T.N. Kaifas, T. Samaras, E. Vafiadis and J.N. Sahalos, "On the Design of Conformal Slot Arrays on Perfectly Conducting Elliptic Cones," *Proc. of the 3rd European Workshop on Conformal Antennas*, pp. 73–76, Bonn (Germany), 2003.

[50] T.N. Kaifas, T. Samaras, K. Siakavara and J.N. Sahalos, "A UTD-OM Technique to Design Slot Arrays on a Perfectly Conducting Paraboloid," *IEEE Trans. Antennas Propag.*, Vol. **AP- 53**, pp. 1688–1698, May 2005.

[51] L.V. Kantorovich and V.I. Krylov, *Approximate Methods for Higher Analysis*, Chapter 1, Interscience Publishers, New York, 1958.

[52] G.A. Thiele, "Wire Antennas," Chapter 2 in *Computer Techniques in Electromagnetics*, Pergamon Press, New York, 1973.

[53] N. Abdelmalek, "Round off Error Analysis for Gram-Schmidt Method and Solution of Linear Least Squares Problems," *BIT*, Vol. **11**, pp. 345–368, 1971.

[54] A. Bjorck, "Numerics of Gram-Schmidt Orthogonalization," *Linear Algebra Appl.*, Vol. **197/198**, pp. 297–316, 1994.

[55] P. Derusso, R. Roy and C. Close, *State Variables for Engineers*, Wiley, New York, 1967.

[56] C.T. Tai, "The Optimum Directivity of Uniformly Spaced Broadside Arrays of Dipoles," *IEEE Trans. Antennas Propag.*, Vol. **AP-12**, pp. 447–454, 1964.

[57] Y.T. Lo, S.W. Lee and Q.H. Lee, "Optimization of Directivity and Signal-to- Noise Ratio of an Arbitrary Antenna Array," *Proc. IEEE*, Vol. **54**, pp. 1033–1045, 1966.

[58] A.T. Moffet, "Minimum Redundancy Linear Arrays", *IEEE Trans. Antennas Propag.*, Vol. **AP-16**, No. 2, pp. 172–175, 1968.

[59] T.H. Ismail and M.M. Dawoud, "Null Steering in Phased Arrays by Controlling the Element Positions", *IEEE Trans. Antennas Propag.*, Vol. **AP-39**, No. 11, pp. 1561–1566, 1991.

[60] B.P. NG, "Designing Array Patterns with Optimum Inter-Element Spacings and Optimum Weights Using a Computer-Aided Approach", *Int. J. Elect.*, Vol. **73**, No. 3, pp. 653–664, 1992.

[61] B.P. NG, M.H. ER, and C. KOT, "Linear Array Geometry Synthesis with Minimum Sidelobe Level and Null Control," *IEE Proc.-Microw. Antennas Propag.*, Vol. **141**, No. 3, pp. 162–166, 1994.

[62] T. Numazaki, S. Mano, T. Kategi and M. Mizusawa, "An Improved Thinning Method for Density Tapering of Planar Array Antennas," *IEEE Trans. Antennas Propag.*, Vol. **AP-35**, pp. 1066–1069, Sept 1987.

[63] R.J. Mailloux and E. Cohen, "Statistically Thinned Arrays with Quantized Element Weights," *IEEE Trans. Antennas Propag.*, Vol. **AP-39**, No. 4, pp. 436–447, April 1991.

[64] R.W.P. King, *The Theory of Linear Antennas*, Harvard University Press, Cambridge, MA, 1956.

[65] J.H. Richmond, "Radiation and Scattering by Thin-wire Structures in the Complex Frequency Domain," *Tech. Report*. 2902-10, The Ohio State University, Electro-Science Laboratory, July 1973.

[66] W.L. Stutzman and G.A. Thiele, *Antenna Theory and Design*, John Wiley & Sons, New York, 1998.

[67] J.N. Sahalos, "Antenna Synthesis by Orthogonal MoM (OM3)," *Proc. of the NATO Advanced Studies in EM*, Samos, 1997.

[68] Z.D. Zaharis, *Antennas and Propagation Modeling for Mobile Communications Systems*, PhD Dissertation, University of Thessaloniki, Greece, (in Greek), 2000.

[69] R.J. Mailloux, *Phased Array Antenna Handbook*, Artech House, Boston, 1994.

[70] H. Steyskal and J.S. Herd, "Mutual Coupling Compensation in Small Array Antennas," *IEEE Trans. Antennas Propag.*, Vol. **38**, pp. 1971–1975, Dec. 1990.

[71] W.H. Kummer, Guest Editor, "Special Issue on Conformal Arrays," *IEEE Trans. Antennas Propag.*, Vol. **AP-22**, No. 1, pp. 1–150, 1974.

[72] O.M. Bucci, G. D'Elia, G. Romito, "Power Synthesis of Conformal Arrays by Generalized Projection Method", *IEE Proc. Microw. Antennas Propag.*, Vol. **142**, No. 6, pp. 467–471, Dec. 1995.

[73] O.M. Bucci and G.D. Elia, "Power Synthesis of Reconfigurable Conformal Array with Phase-Only Control," *IEE Proc.-Microw. Antennas Propag.*, Vol. **145**, No. 1, pp. 131–135, Feb. 1998.

[74] O. Schmid, "*Pattern Synthesis for Large Conformal Array Using Two-Port Element for Polarimetric Correction*," PIERS 98, Nantes, (France), 1998.

[75] J.C. Sureau and K.J. Keeping, "Sidelobe Control in Cylindrical Arrays," *IEEE Trans. Antennas Propag.*, Vol. **AP-30**, No. 5, pp. 1027–1031, Sept. 1982.

[76] E.C. Duford, "Pattern Synthesis Based on Adaptive Array Theory," *IEEE Trans. Antennas Propag.*, Vol. **AP-37**, pp. 1017–1018, 1989.

[77] C.A. Olsen and R.T. Compton Jr, "A Numerical Pattern Synthesis Algorithm for Arrays," *IEEE Trans. Antennas Propag.*, Vol. **AP-38**, No. 10, pp. 1666–1676, Oct. 1990.

[78] P. Pathak, N. Wang, W.D. Burnside and R. Kouyoumjian, "A Uniform GTD Solution for the Radiation from Sources on a Convex Surface," *IEEE Trans. Antennas Propag.*, Vol. **29**, No. 4, pp. 609–622, 1981.

[79] P. Pathak and N. Wang, "Ray Analysis of Mutual Coupling between Antennas on a Convex Surface," *IEEE Trans. Antennas Propag.*, Vol. **29**, No. 6, pp. 911–922, 1981.

[80] S. Blume and U. Usckerat, "The Radar Cross-Section of the Semiinfinite Elliptic Cone: Numerical Evaluation," *Wave Motion*, Vol. **22**, pp. 311–324, 1995.

[81] S. Blume, L. Klinkenbusch and U. Usckerat, "The Radar Cross-Section of the Semiinfinite Elliptic Cone," *Wave Motion*, Vol. **17**, pp. 365–389, 1993.

[82] S. Blume and V. Krebs, "Numerical Evaluation of Dyadic Diffraction Coefficients and Bistatic Radar Cross Sections for a Perfectly Conducting Semi-infinite Elliptic Cone," *IEEE Trans. Antennas Propag.*, Vol. **46**, No. 3, pp. 414–424, 1998.

[83] E. Vafiadis and J.N. Sahalos, "The Electromagnetic Field of a Slotted Elliptic Cone," *IEEE Trans. Antennas Propag.*, Vol. **38**, No. 11, pp. 1894–1890, 1990.

[84] C.-T. Tai, *Dyadic Green's Functions in Electromagnetic Theory*, IEEE Press, New York, 1994.

[85] J.R. James and P.S. Hall, *Handbook of Microstrip Antennas*, Peter Peregrinus Ltd, London, 1989.

[86] K.R. Carver and J.W. Mink, "Microstrip Antenna Technology," *IEEE Trans. Antennas Propag.*, Vol. **29**, pp. 2–24, 1981.

[87] A.T. Villeneuve, M.C. Behnke and W.H. Kummer, "Wide-Angle Scanning of Linear Arrays Located on Cones," *IEEE Trans. Antennas Propag.*, Vol. **22**, No. 1, pp. 97–103, Jan. 1974.

[88] G.A. Thiele and C. Donn, "Design of a Small Conformal Array," *IEEE Trans. Antennas Propag.*, Vol. **22**, No. 1, pp. 64–70, Jan. 1974.

[89] Q. Balzano and T. Dowling, "Mutual Coupling Analysis of Arrays of Apertures on Cones," *IEEE Trans. Antennas Propag.*, Vol. **22**, No. 1, pp. 92–97, Jan. 1974.

[90] F.G. Hansen, *Phased Array Antennas*, John Wiley & Sons, New York, 1998.

[91] D.M. Pozar, *Microwave Engineering*, John Wiley & Sons, New York, 1997.

[92] M.M. Lipschutz, *Differential Geometry*, McGraw-Hill, New York, 1995.

[93] A. Michaeli, "Equivalent Edge Currents for Arbitrary Aspects of Observation," *IEEE Trans. Antennas Propag.*, Vol. **32**, No. 3, pp. 252–258, 1984.

5

The ORAMA Computer Tool

George S. Miaris and John N. Sahalos

5.1 Introduction

In this chapter, the Orthogonal Advanced Methods for Antennas (ORAMA) computer tool will be presented. ORAMA has been designed to meet the needs of post-graduate students and professional antenna engineers.

ORAMA is based on the orthogonal method (OM) and has been designed for the element excitation derivation of linear antenna arrays. OR.A.M.A comes from the initials of the words **OR**thogonal **A**dvanced **M**ethods for **A**ntennas. ORAMA is a Greek word that means 'vision'. This tool has been written through a menu driven form to be user-friendly. Mutual coupling between the elements of the arrays is not taken into account in the present version. However, ORAMA can make excellent pattern predictions for different element types.

The elements through the phase and amplitude control can vary across the aperture of the array and allow the realization of certain requirements related to pattern characteristics.

The antenna array is designed for different types and positions of elements depending on needs. Elements such as dipoles, microstrips and horns are common in most of the applications.

In this program, parallel and identical elements are taken into account.

5.2 Description of the ORAMA Program

ORAMA is written for the Windows 98/NT/2000/XP operating systems. It has been designed as a Windows MDI application so that multiple array designs can be performed simultaneously. The program starts with the screen of Fig. 5.1.

There are four tabs:

1. Array data
2. Excitation coefficients
3. Main plot
4. Plots.

Orthogonal Methods for Array Synthesis: Theory and the ORAMA Computer Tool   John N. Sahalos
© 2006 John Wiley & Sons, Ltd

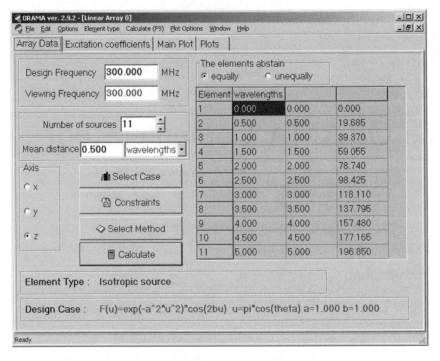

Figure 5.1 The ORAMA start-up screen

The array data tab can be used to enter all the data that are necessary for the description of the antenna. The user has the following options:

1. Set the design frequency of the array. This frequency is used to convert lengths from wavelengths into metres or inches and vice versa.
2. Set the viewing frequency of the array. This frequency is used to view the radiation pattern of the designed array at a different frequency.
3. Set the number of sources.
4. Set the inter-element distance. This edit box is disabled if elements are at unequal distances. To view or change mean distance in metres or inches the user can select the appropriate unit from the drop-down list box on the right side of the edit box.
5. Choose the array axis x, y or z.
6. By clicking the button 'Select Case' or by pressing the 'F7' key, or by selecting the menu 'Options | Design Case...', a dialog box will appear from which the user can choose one of the built-in design cases. The cases have been selected to provide a great variety of desired factors, and have been parameterized to allow additional degrees of freedom. Figure 5.2 shows the desired array factors, which have to do with most of the practical design cases. For example, by clicking the Chebyshev/Legendre case, a variety of the above patterns (see Fig. 5.3) will be presented. By clicking one of the patterns (see Fig. 5.4), the desired details with the corresponding constraints will be presented.

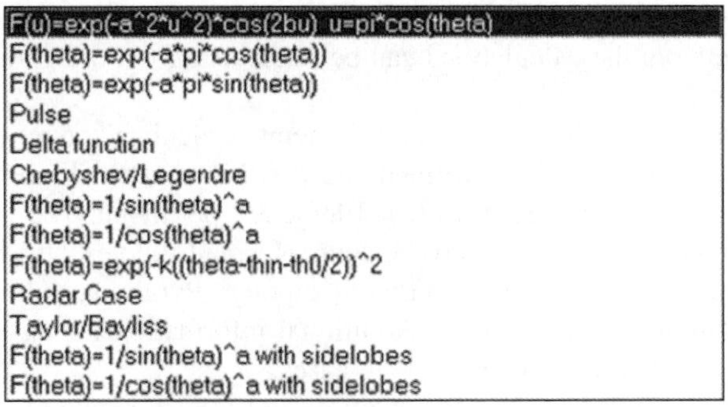
F(u)=exp(-a^2*u^2)*cos(2bu) u=pi*cos(theta)
F(theta)=exp(-a*pi*cos(theta))
F(theta)=exp(-a*pi*sin(theta))
Pulse
Delta function
Chebyshev/Legendre
F(theta)=1/sin(theta)^a
F(theta)=1/cos(theta)^a
F(theta)=exp(-k((theta-thin-th0/2))^2
Radar Case
Taylor/Bayliss
F(theta)=1/sin(theta)^a with sidelobes
F(theta)=1/cos(theta)^a with sidelobes

Figure 5.2 The ORAMA design cases list box

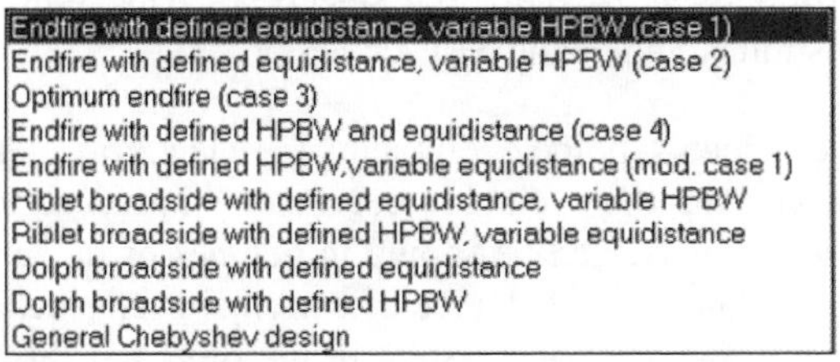
Endfire with defined equidistance, variable HPBW (case 1)
Endfire with defined equidistance, variable HPBW (case 2)
Optimum endfire (case 3)
Endfire with defined HPBW and equidistance (case 4)
Endfire with defined HPBW,variable equidistance (mod. case 1)
Riblet broadside with defined equidistance, variable HPBW
Riblet broadside with defined HPBW, variable equidistance
Dolph broadside with defined equidistance
Dolph broadside with defined HPBW
General Chebyshev design

Figure 5.3 The ORAMA Chebyshev/Legendre design cases list box

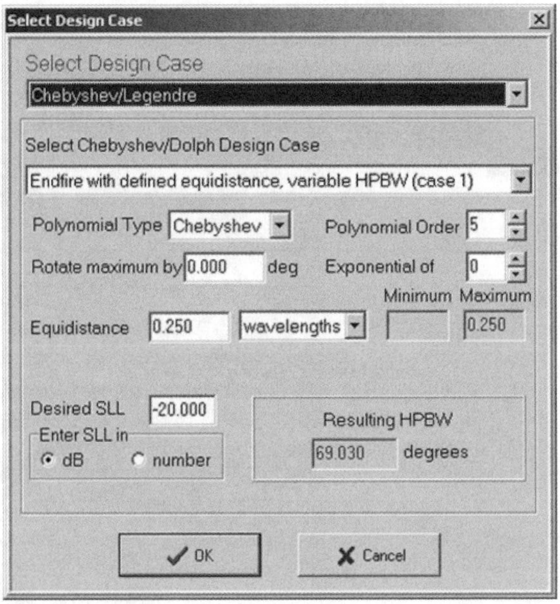
Select Design Case
Select Design Case
Chebyshev/Legendre
Select Chebyshev/Dolph Design Case
Endfire with defined equidistance, variable HPBW (case 1)
Polynomial Type
Chebyshev
Polynomial Order
5
Rotate maximum by
0.000
deg
Exponential of
0
Minimum
Maximum
Equidistance
0.250
wavelengths
0.250
Desired SLL
-20.000
Enter SLL in
dB
number
Resulting HPBW
69.030
degrees
OK
Cancel

Figure 5.4 The ORAMA Chebyshev/Legendre case selection dialog box

7. The button 'Select Method' is used for the selection of the method of calculation. In the present version, the calculations can be made using the orthogonal method given in paragraph 4.3.
8. By clicking the button 'Calculate' or the menu 'Calculate' or by pressing the 'F9' key, the calculations will be performed and the 'Plots' tab will appear.
9. The elements can be selected from the 'Element type' menu. The following element types are currently supported: isotropic sources, small dipoles, small crossed dipoles, finite length dipoles, finite length crossed dipoles, Pyramidal horns and microstrip patches (Rectangular and Circular). Additional information for element type characteristics is required in most of the above cases.
10. The use of a planar reflector in the array design is also supported. From the Menu 'Options' the user selects 'Reflector'. Then, a dialog box appears and the user has to choose the reflector position.
11. The button 'Constraints' is used for the selection of the null constraints, the area minimization constraints and/or the SLL constraints.

In the excitation coefficients tab, the coefficients for each array element are shown. The user has the option of viewing the excitation coefficients in rectangular or polar form. In the case of polar form, the phase can be shown in degrees or radians.

When the checkbox 'Allow manual change of currents' is checked, the user can change the excitation coefficients and view the results in the plots tab. In the dialog box, there is also the possibility of choosing a pre-selected excitation distribution. In Fig. 5.5, the pre-selection dialog box is presented.

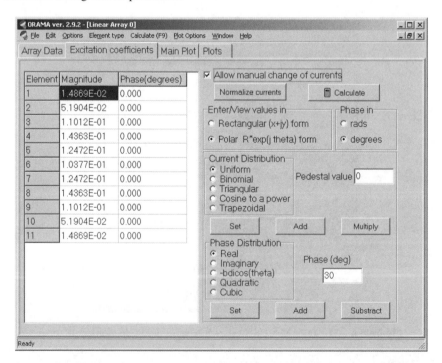

Figure 5.5   The ORAMA dialog box of the pre-selection of the excitation distribution

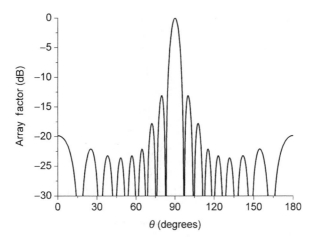

Figure 5.6   Pattern of a linear array with $N = 11, d/\lambda = 0.75$ and $AF(\theta) = \delta(\theta - 90°)$.

It is noticed that, besides setting, the pre-selected excitation can be added or multiplied to the one found by the OM of a desired pattern. If a pattern taken by the OM has side lobes higher than the desired ones, an addition or multiplication of the calculated excitation with a pre-selected one could give lower SLL. Let us suppose that a linear array with $N = 11$ elements and $d/\lambda = 0.75$ is used to have an array factor $AF(\theta) = \delta(\theta - 90°)$. By applying the OM, a pattern with SLL $\cong -13$ dB is achieved (see Fig. 5.6).

If it is desirable to lower SLL, then we can add a triangular pre-selected excitation. In this case, the SLL becomes -16.7 dB, (Fig. 5.7).

Moreover, if it is desirable to lower the SLL even more, we can add the triangular excitation again and have SLL $= -19.2$ dB, (Fig. 5.8).

The procedure can be continued until the desired SLL is achieved. In Table 5.1, the array excitations for the three different cases are presented.

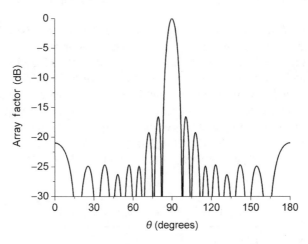

Figure 5.7   Pattern of a linear array with $N = 11, d/\lambda = 0.75$ and $AF(\theta) = \delta(\theta - 90°)$, where a triangular excitation distribution is added

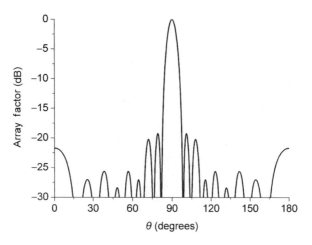

Figure 5.8   Pattern of a linear array with $N = 11$, $d/\lambda = 0.75$ and $AF(\theta) = \delta(\theta - 90°)$, where a triangular excitation distribution is added twice

Table 5.1   The array excitations for the patterns of Figs. 5.6, 5.7 and 5.8

Element no	1,11	2,10	3,9	4,8	5,7	6
A_i (Fig. 5.6)	0.826	0.995	1.0	0.953	0.961	0.982
A_i (Fig. 5.7)	0.573	0.744	0.813	0.850	0.921	1.0
A_i (Fig. 5.8)	0.458	0.627	0.724	0.798	0.896	1.0

Another option to lower the SLL is by the multiplication of the excitations with the pre-selected ones. In this case, the SLL becomes -23.8 dB. Figure 5.9 shows the pattern achieved and Table 5.2 presents the final excitation of the array. It is noticed that instead of the triangular distribution one could choose another one on the menu and compare the results.

If it is desirable to rotate the main lobe, a phase difference can be added through the dialog box. In order to have a rotation of $10°$ for the linear array given above, we use the key '-bdcos(θ)' for $\theta = 80°$. So, with the excitation of Table 5.2 and a phase difference between the elements equal to $46.9°$, we have the pattern of Fig. 5.10.

With the key 'imaginary' in the 'phase distribution' box, one can increase or decrease exponentially the amplitudes of the element excitations. For a 'damping factor' equal to -0.3, the pattern of Fig. 5.6 is modified to that of Fig. 5.11, which has lower SLL.

On the 'Main Plot' tab (Fig. 5.12), the pattern plots are shown enlarged, one plane at a time. The user has the option of customizing the plot view. The description of the most significant options follows.

Table 5.2   The array excitations for the pattern of Fig. 5.9

Element no	1,11	2,10	3,9	4,8	5,7	6
A_i	0.140	0.338	0.509	0.647	0.815	1.0

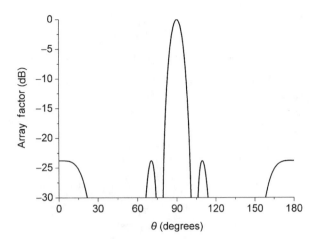

Figure 5.9 Pattern of a linear array with $N = 11, d/\lambda = 0.75$ and $AF(\theta) = \delta(\theta - 90°)$, where a triangular excitation distribution is multiplied

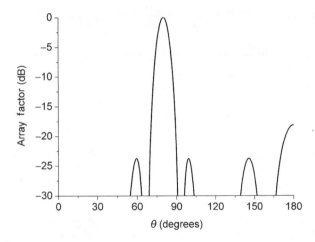

Figure 5.10 Pattern of a linear array with $N = 11, d/\lambda = 0.75$ and initial $AF(\theta) = \delta(\theta - 90°)$, where a triangular excitation distribution is multiplied and a phase difference of $46.9°$ is added

The plane of the plot can be selected or even entered through the angle ϕ. The option of plotting the total factor, the element or the array factor is also possible. The graph can be viewed in either polar or rectangular form. The graph can be zoomed in by checking the 'Edit Plot' checkbox and by selecting the area to zoom in with the mouse.

On the 'Plots' tab, the array patterns and the array indices at four planes are shown. The planes shown (Fig. 5.13) on the 'Plots' tab are xy, yz, xz and a constant ϕ cut on a user definable plane (the default is $\phi = 45°$).

A 3-D graph option (Fig. 5.14) is also possible by selecting '3-D Graph' from the 'Graph' menu.

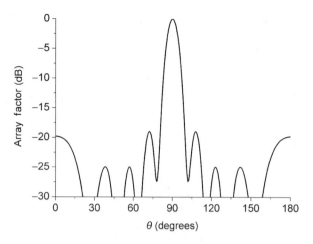

Figure 5.11 Pattern of a linear array with $N = 11$, $d/\lambda = 0.75$ and initial $AF(\theta) = \delta(\theta - 90°)$, where an imaginary phase difference with -0.3 damping factor is added

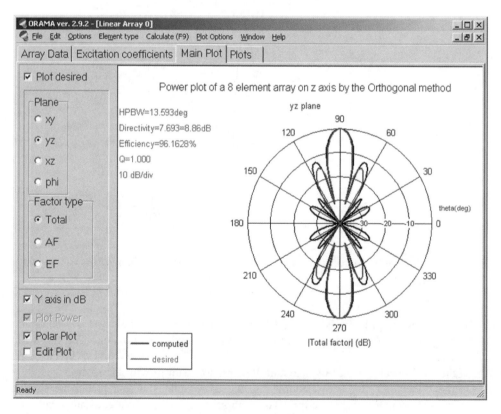

Figure 5.12 The main plot tab

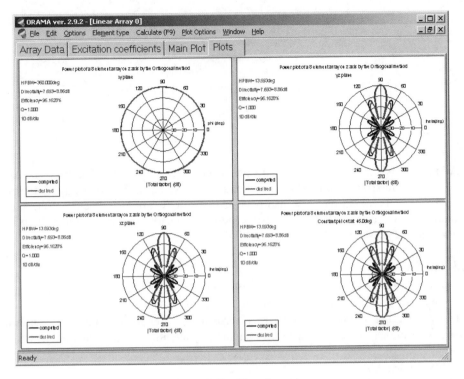

Figure 5.13   The plots tab

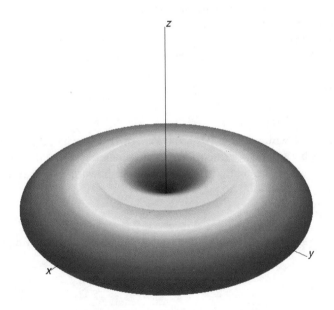

Figure 5.14   A 3-D pattern example

Figure 5.15   The file menu

Additional useful operations exist in the menu 'File' (Fig. 5.15). The user can save, open or close an array design. The main plot can be saved as a Windows bitmap file. The pattern data of the main plot can also be saved in ASCII format. The current array data in the main plot can be printed on every Windows compatible printer through the menus 'Print Numeric Results' and 'Print Main Plot'.

5.3 Element Types

Elements can be selected by using the menu item 'Element type'. In this paragraph, the elements used in ORAMA will be presented.

It is noticed that in synthesis, ORAMA makes use of the array factor. Also, ORAMA does not monitor element size. The user should pay attention to array and element size.

5.3.1 Isotropic Source

An element having equal radiation in all directions is called an 'isotropic source'. Although it is physically unrealizable, the isotropic source is frequently used in the array design process because it helps in the exploitation of array factor characteristics.

An isotropic source is a point source that can be simulated by a delta function excitation [1]. It is easy to conclude that the normalized element factor for an isotropic source is

$$f(\theta, \phi) = 1 \tag{5-1}$$

5.3.2 Small Dipole (Hertz Dipole)

An infinitesimal wire with a length $l \leq \lambda/50$ is known as a 'small dipole'. Following [1] and [2], it is possible to obtain closed form expressions for the field radiated by a small dipole. By imposing the far field condition, it can be shown that the normalized element factor of a z directed small dipole is

$$f(\theta, \phi) = \sin \theta \tag{5-2}$$

Likewise, the normalized element factors for x and y directed small dipoles are

$$f(\theta, \phi) = \sqrt{1 - \cos^2 \phi \sin^2 \theta} \tag{5-3}$$

and

$$f(\theta, \phi) = \sqrt{1 - \sin^2 \phi \sin^2 \theta} \tag{5-4}$$

respectively.

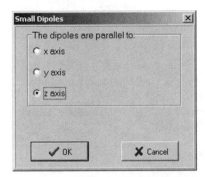

Figure 5.16 Dialog box specifying the small dipoles direction

When the 'small dipole' is selected as the array element, a dialog box opens (see Fig.5.16) asking for the dipoles direction. The dipoles can be parallel to the x, y or z-axis.

5.3.3 Finite Length Dipole

Finite length dipoles are widely used as simple antennas and as radiating elements in arrays. The geometry of a finite length dipole can be seen in Fig. 5.17.

The full analysis of obtaining the element factor of a finite length dipole is beyond the scope of this book. Therefore, only the results will be presented. For more information, the reader is referred to [1, 2].

It is assumed that the current distribution on a z directed dipole having length $2L$ is given by

$$I(z) = I_m \sin[\beta(L - |z|)] \tag{5-5}$$

I_m is the maximum of the current.

It can be shown that the normalized element factor is:

$$f(\theta, \phi) = \frac{\cos(\beta L \cos \theta) - \cos(\beta L)}{\sin \theta} \tag{5-6}$$

When the dipole is on the x or y-axis, the corresponding factors are

$$f(\theta, \phi) = \frac{\cos(\beta L \cos \phi \sin \theta) - \cos(\beta L)}{\sqrt{1 - \cos^2 \phi \sin^2 \theta}} \tag{5-7}$$

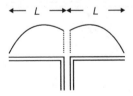

Figure 5.17 A finite length dipole with length $2L$ and a sinusoidal current distribution

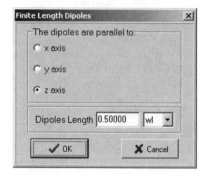

Figure 5.18   Dialog box specifying the dipoles direction and length

and

$$f(\theta, \phi) = \frac{\cos(\beta L \sin \phi \sin \theta) - \cos(\beta L)}{\sqrt{1 - \sin^2 \phi \sin^2 \theta}} \tag{5-8}$$

respectively.

When the 'finite length dipole' is selected as the array element, a dialog box opens (see Fig.5.18) asking for the direction and length of the dipoles. The dipoles can be parallel to the x, y or z-axis. The length of the dipoles is derived in wavelengths, metres or inches.

5.3.4 Crossed Dipoles

Two-crossed dipoles (see Fig. 5.19), perpendicular to each other, create an almost omni-directional element. It can be shown [3], that the input currents are $I_1 = 1$ and $I_2 = 1 \lfloor -\pi/2$ for the first and second dipole respectively.

The element factor of the two-crossed dipoles is of the form

$$f(\theta, \phi) = f_1(\theta, \phi) + j f_2(\theta, \phi) \tag{5-9}$$

where $f_1(\theta, \phi)$ and $f_2(\theta, \phi)$ are the element factors of the first and second dipole respectively.

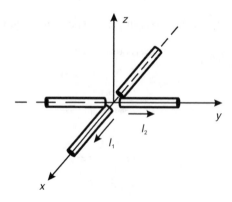

Figure 5.19   Geometry of two-crossed dipoles on the xy-plane

Figure 5.20   Dialog box specifying the plane of the crossed dipoles

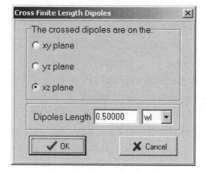

Figure 5.21   Dialog box specifying the crossed dipoles plane and length

The element factor of two-small crossed dipoles can be expressed in the form

$$f(\theta, \phi) = \sqrt{\frac{1 + u^2(\theta, \phi)}{2}} \tag{5-10}$$

where

$$u(\theta, \phi) = \begin{cases} \cos\theta, & \text{for a crossed dipole on the } xy \text{ plane} \\ \cos\phi\sin\theta, & \text{for a crossed dipole on the } yz \text{ plane} \\ \sin\phi\sin\theta, & \text{for a crossed dipole on the } xz \text{ plane} \end{cases} \tag{5-11}$$

Finite length crossed dipoles are calculated in ORAMA by combining Equations (5-6)–(5-9).

When the 'crossed small dipole' or the 'crossed finite length dipole' is selected as the array element, a dialog box opens (see Fig. 5.20 or 5.21), asking for each dipole's direction and/or length. The dipoles can be on the xy, yz or xz plane. The length of the dipoles is derived in wavelengths, metres or inches.

5.3.5 Horns

Horns are widely used in the GHz region. They are a special case of the so-called 'aperture antennas' [4]. Next, the formulation for the rectangular and the sectoral horns with smooth or corrugated walls will be given.

We define two vectors $\boldsymbol{P}$ and $\boldsymbol{Q}$ as the two-dimensional Fourier transforms of the equivalent $\boldsymbol{E}_s$ and $\boldsymbol{H}_s$ fields on the bounding plane of the horn. It is assumed that the aperture is on the xy plane. The components of the vectors are

$$P_x = \iint\limits_S \boldsymbol{E}_s \cdot \hat{\boldsymbol{x}} e^{j\beta(x' \sin\theta \cos\phi + y' \sin\theta \sin\phi)} \, dx' \, dy' \tag{5-12}$$

$$P_y = \iint\limits_S \boldsymbol{E}_s \cdot \hat{\boldsymbol{y}} e^{j\beta b(x' \sin\theta \cos\phi + y' \sin\theta \sin\phi)} \, dx' \, dy' \tag{5-13}$$

$$Q_x = \iint\limits_S \boldsymbol{H}_s \cdot \hat{\boldsymbol{x}} e^{j\beta b(x' \sin\theta \cos\phi + y' \sin\theta \sin\phi)} \, dx' \, dy' \tag{5-14}$$

$$Q_y = \iint\limits_S \boldsymbol{H}_s \cdot \hat{\boldsymbol{y}} e^{j\beta b(x' \sin\theta \cos\phi + y' \sin\theta \sin\phi)} \, dx' \, dy' \tag{5-15}$$

Calculations can be greatly simplified if only one type of field is used. As a consequence, three different and equivalent formulations exist:

- When the tangential components of the magnetic and electric fields over a closed surface passing through the aperture are used, the components of the far field are

$$E_\theta = j\beta \frac{e^{-j\beta r}}{4\pi r} [P_x \cos\phi + P_y \sin\phi + n \cos\theta (Q_y \cos\phi - Q_x \sin\phi)] \tag{5-16}$$

$$E_\phi = j\beta \frac{e^{-j\beta r}}{4\pi r} [\cos\theta (P_y \cos\phi - P_x \sin\phi) - n(Q_y \sin\phi + Q_x \cos\phi)] \tag{5-17}$$

- When only the magnetic field on the aperture is used, the model is referred to as the 'H-field model'. The far field is

$$E_\theta = j\beta n \frac{e^{-j\beta r}}{2\pi r} \cos\theta (Q_y \cos\phi - Q_x \sin\phi) \tag{5-18}$$

$$E_\phi = -j\beta n \frac{e^{-j\beta r}}{2\pi r} (Q_y \sin\phi + Q_x \cos\phi) \tag{5-19}$$

- When only the electric field on the aperture is used, the model is referred to as the 'E-field model'. The far field is

$$E_\theta = j\beta \frac{e^{-j\beta r}}{2\pi r} (P_x \cos\phi + P_y \sin\phi) \tag{5-20}$$

$$E_\phi = j\beta \frac{e^{-j\beta r}}{2\pi r} \cos\theta (P_y \cos\phi - P_x \sin\phi) \tag{5-21}$$

It has been found [5] that the E-field model produces results that better match the measurements. So, following [4] this model is used in ORAMA.

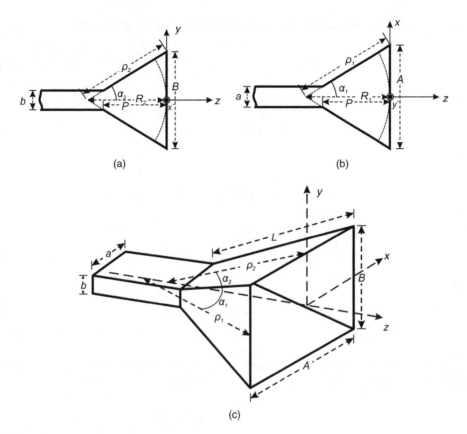

Figure 5.22 Geometry of a pyramidal horn (a) E-plane geometry (b) H-plane geometry (c) Horn overview

5.3.5.1 Pyramidal Horns

The geometry of a smooth-walled pyramidal horn can be seen in Fig. 5.22, [2, 6]. The parameters of this horn are the aperture dimensions A, B, the semi-flare angles α_1 and α_2, the distances from the aperture to the horn apex R_1, R_2 and the distance from the waveguide aperture to the horn aperture P. The plane parallel to the greater waveguide dimension is referred to as the 'H-plane' while the one parallel to the smaller waveguide dimension is the 'E-plane'.

It is necessary to derive the electric field on the aperture in order to calculate the far field using the E-model. It is assumed that the TE_{10} mode exists in the rectangular waveguide.

The field at the horn aperture is

$$E_S = \hat{y} E_0 \cos \frac{\pi x}{A} e^{-j\frac{\beta}{2}\left(\frac{x^2}{R_1}+\frac{y^2}{R_2}\right)}$$ (5-22)

Equation (5-22) implies that the field is linearly polarized in the y direction. When circular polarization is desired, an equal amplitude with a $90°$ phase shift of the TE_{01}

mode is excited in the rectangular waveguide. In this case, the aperture field is

$$\mathbf{E}_S = E_0 \left(\hat{\mathbf{y}} \cos \frac{\pi x}{A} + j\hat{\mathbf{x}} \cos \frac{\pi y}{B} \right) e^{-j\frac{\beta}{2}\left(\frac{x^2}{R_1} + \frac{y^2}{R_2} \right)} \tag{5-23}$$

Substituting (5-22) to (5-13) one has

$$\mathbf{P} = P_y \hat{\mathbf{y}} = E_0 I_{1x} I_{2y} \tag{5-24}$$

where

$$I_{1x} = \int_{-A/2}^{A/2} \cos \left(\frac{\pi x}{A} \right) e^{-j\beta\left(\frac{x^2}{2R_1} - ux \right)} dx \tag{5-25}$$

$$I_{2y} = \int_{-B/2}^{B/2} e^{-j\beta\left(\frac{y^2}{R_2} - vy \right)} dy \tag{5-26}$$

and

$$u = \sin \theta \cos \phi \tag{5-27}$$

$$v = \sin \theta \sin \phi \tag{5-28}$$

If $\alpha_1 = 0$ or $\alpha_2 = 0$, the horn is called 'E-plane' or 'H-plane sectoral horn' respectively. When $\alpha_1 = 0$, then R_1 tends to infinity and (5-25) yields

$$I_{1x} = - \left(\frac{\pi A}{2} \right) \frac{\cos \left(\frac{\beta A u}{2} \right)}{(\beta A u/2)^2 - (\pi/2)^2} \tag{5-29}$$

When $\alpha_2 = 0$, then R_2 tends to infinity and (5-26) becomes

$$I_{2y} = B \frac{\sin(\beta B v/2)}{\beta B v/2} \tag{5-30}$$

For circular polarization, it is

$$\mathbf{P} = P_x \hat{\mathbf{x}} + j P_y \hat{\mathbf{y}} \tag{5-31}$$

P_x and P_y are calculated from Equations (5-12) and (5-13) respectively. The integrals can be expressed in terms of cosine and sine Fresnel integrals [7,8]:

$$C(x) = \int_0^x \cos \left(\frac{\pi}{2} t^2 \right) dt \tag{5-32}$$

$$S(x) = \int_0^x \sin \left(\frac{\pi}{2} t^2 \right) dt \tag{5-33}$$

By defining

$$F(x) = C(x) + j S(x) \tag{5-34}$$

P_x can be expressed as

$$P_x = jE_0\frac{1}{2}\sqrt{\frac{R_2\pi}{\beta}}\left[e^{j\frac{\beta R_2}{2}(v+(\pi|\beta B))^2}(F^*(S_2') - F^*(S_1')) + e^{j\frac{\beta R_2}{2}(v-(\pi|\beta B))^2}(F^*(t_2')\right.$$

$$\left. - F^*(t_1'))\right] \cdot \sqrt{\frac{R_1\pi}{\beta}}\left[e^{j\frac{\beta R_1}{2}u^2}(F^*(p_2') - F^*(p_1'))\right] \tag{5-35}$$

where

$$\left.\begin{aligned}
p_2' &= \sqrt{\beta/\pi R_1}(A/2 - R_1 u)\\
p_1' &= \sqrt{\beta/\pi R_1}(-A/2 - R_1 u)\\
t_2' &= \sqrt{\beta/\pi R_2}\left(B/2 - R_2 v + \frac{\pi R_2}{\beta B}\right)\\
t_1' &= \sqrt{\beta/\pi R_2}\left(-B/2 - R_2 v + \frac{\pi R_2}{\beta B}\right)\\
s_2' &= \sqrt{\beta/\pi R_2}\left(B/2 - R_2 v - \frac{\pi R_2}{\beta B}\right)\\
s_1' &= \sqrt{\beta/\pi R_2}\left(-B/2 - R_2 v - \frac{\pi R_2}{\beta B}\right)
\end{aligned}\right\} \tag{5-36}$$

Also, P_y is expressed as

$$P_y = E_0\frac{1}{2}\sqrt{\frac{R_1\pi}{\beta}}\left[e^{j\frac{\beta R_1}{2}(u+(\pi|\beta A))^2}(F^*(S_2) - F^*(S_1)) + e^{j\frac{\beta R_1}{2}(u-(\pi|\beta A))^2}(F^*(t_2)\right.$$

$$\left. - F^*(t_1))\right] \cdot \sqrt{\frac{R_2\pi}{\beta}}\left[e^{j\frac{\beta R_2}{2}v^2}(F^*(p_2) - F^*(p_1))\right] \tag{5-37}$$

where

$$\left.\begin{aligned}
p_2 &= \sqrt{\beta/\pi R_2}(B/2 - R_{12} v)\\
p_1 &= \sqrt{\beta/\pi R_2}(-B/2 - R_2 v)\\
t_2 &= \sqrt{\beta/\pi R_1}\left(A/2 - R_1 u + \frac{\pi R_1}{\beta A}\right)\\
t_1 &= \sqrt{\beta/\pi R_1}\left(-A/2 - R_1 u + \frac{\pi R_1}{\beta A}\right)\\
s_2 &= \sqrt{\beta/\pi R_1}\left(A/2 - R_1 u - \frac{\pi R_1}{\beta A}\right)\\
s_1 &= \sqrt{\beta/\pi R_1}\left(-A/2 - R_1 u - \frac{\pi R_1}{\beta A}\right)
\end{aligned}\right\} \tag{5-38}$$

In many cases, it is desirable to design optimal horns. The word 'optimal' characterizes a horn that delivers a specified gain with the smallest possible aperture dimensions.

5.3.5.2 Pyramidal Horn Design

Pyramidal horn optimum design can be made under or without constraints on half-power beamwidths (HPBWs). Three separate cases are the most interesting ones. One derives the horn optimum dimensions for certain desired directivity. The other gives the horn dimensions under constraints on HPBWs on the two main planes (E and H). Finally, the third one derives the horn dimensions under constraints on the ratio of the HPBWs on the two main planes and the directivity.

By taking into account the geometry given in Fig. 5.22, we derive the quadratic phase distribution constants t and s [2]:

$$t = \frac{1}{8}\left(\frac{A}{\lambda}\right)^2 \frac{1}{R_1/\lambda} = \frac{A(A-a)}{8\lambda P} \tag{5-39}$$

$$s = \frac{1}{8}\left(\frac{B}{\lambda}\right)^2 \frac{1}{R_2/\lambda} = \frac{B(B-b)}{8\lambda P} \tag{5-40}$$

The directivity of the antenna in closed form expression is

$$D_{dB} = 10\log\left[\frac{\pi}{8}\frac{AB}{\lambda^2}\frac{X(s)}{s}\frac{Y(t)}{t}\right] \tag{5-41}$$

where

$$X(s) = C^2(2\sqrt{s}) + S^2(2\sqrt{s}) \tag{5-42}$$

$$Y(t) = \left[C\left(\frac{1}{4\sqrt{t}}+2\sqrt{t}\right) - C\left(\frac{1}{4\sqrt{t}}-2\sqrt{t}\right)\right]^2 + \left[S\left(\frac{1}{4\sqrt{t}}+2\sqrt{t}\right)\right.$$
$$\left. - S\left(\frac{1}{4\sqrt{t}}-2\sqrt{t}\right)\right]^2 \tag{5-43}$$

The optimum unconstrained directivity occurs [1–2] when $t = t_{opt} = 0.375$ and $s = s_{opt} = 0.25$.

For $t \neq t_{opt}$ and $s \neq s_{opt}$, directivity becomes smaller than the corresponding optimum one. There are however, cases for which HPBWs on the two main planes or their ratio are just as important as directivity maximization. To design a horn antenna with certain HPBWs, we have to know a relation between s, t and the corresponding HPBWs. By applying fitting techniques for a large amount of horns, two approximate expressions can be used [9]. These are:

$$\frac{A}{\lambda} \cdot \frac{\tan\left(\frac{HP_H}{2}\right)}{\cos\left(\frac{HP_H}{2}\right)} = \frac{1}{2}\sqrt{\frac{a_H + c_H t^2 + e_H t^4 + g_H t^6 + i_H t^8}{1 + b_H t^2 + d_H t^4 + f_H t^6 + h_H t^8}} \qquad t \leq 0.88 \tag{5-44}$$

$$\frac{B}{\lambda} \cdot \frac{\tan\left(\frac{HP_E}{2}\right)}{\cos\left(\frac{HP_E}{2}\right)} = \frac{1}{2}\sqrt{\frac{a_E + c_E s^2 + e_E s^4 + g_E s^6}{1 + b_E s^2 + d_E s^4 + f_E s^6 + h_E s^8}} \qquad s \leq 0.47 \tag{5-45}$$

The coefficients involved in Equations (5-44) and (5-45) are given in Tables 5.3 and 5.4 respectively.

Table 5.3 Coefficients of Equation (5-44)

a_H	b_H	c_H	d_H	e_H	f_H	g_H	h_H	i_H
0.3534	−5.9711	−1.5379	13.4735	2.4359	−13.3102	−1.6386	4.7981	0.6333

Table 5.4 Coefficients of Equation (5-45)

a_E	b_E	c_E	d_E	e_E	f_E	g_E	h_E
0.1962	−11.3448	−1.9135	41.7800	5.7284	−47.9711	−4.8935	−6.4175

Let us suppose that it is desirable to design a horn antenna with HPBWs, which on the E- and H-planes are the HP_E and HP_H respectively. For a given HP_E one can find an $HP_H^o \neq HP_H$ corresponding to the optimum without constraints horn, that is, the one that has $s = 0.25$ and $t = 0.375$. In addition, for a given HP_H, an $HP_E^o \neq HP_E$ corresponding to the horn with $s = 0.25$ and $t = 0.375$ can be derived.

The design procedure starts with the derivation of the following quantities: $k = \dfrac{HP_H}{HP_E}$, $k_{opt}^E = \dfrac{HP_H^o}{HP_E}$ and $k_{opt}^H = \dfrac{HP_H}{HP_E^o}$. HP_H, HP_E, HP_H^o and HP_E^o are measured in degrees. In order to design the optimum horn under constraints on the HPBWs, one of the parameters s or t has to retain the value $s = 0.25$ or $t = 0.375$ of the optimum horn while the other parameter changes.

SET 1.

For $k \leq \min(k_{opt}^E, k_{opt}^H)$, t must be equal to 0.375 while constant s becomes

$$s = \left(a_0 + \frac{a_1}{k^2} + \frac{a_2}{D_{dB}} \right)^{-1} \tag{5-46}$$

Directivity is calculated from [9]

$$D_{dB} = c_0(k) + c_1(k)\ln(HP_H) \tag{5-47}$$

where

$$c_0(k) = c_{00} + c_{01}\ln(k) \tag{5-48}$$

$$c_1(k) = c_{10} + c_{11}k + c_{12}k^2 + c_{13}k^3 + c_{14}k^4 + c_{15}k^5 \tag{5-49}$$

Table 5.5 Coefficients of Equation (5-46)

a_0	a_1	a_2
0.2974	7.0401	−37.5383

Table 5.6 Coefficients of Equations (5-48) and (5-49)

c_{00}	c_{01}	c_{10}	c_{11}	c_{12}	c_{13}	c_{14}	c_{15}
44.8365	4.3374	−8.1501	−2.9183	8.4217	−13.2623	10.6702	−3.4713

The design procedure can be summarized as follows: First, from (5-47) D_{dB} is calculated. Then, from (5-46) s is found. Knowledge of parameters t and s and HPBW in the two principal planes gives A/λ and B/λ from (5-44) and (5-45). Finally, the horn length R is calculated from (5-39).

The results do not give the exact outcome. So, in order to improve the design an iterative procedure between Equations (5-40) and (5-46) can be followed. After calculating s from (5-40), an updated value for B/λ can be found from (5-46) by using the fixed value of R found from (5-39). The process is repeated until the correction to B/λ becomes smaller than a given tolerance ϵ. Usually, two or three iterations are enough for $\epsilon \leq 10^{-4}$.

SET 2.

For $k \geq \max(k_{opt}^{E}, k_{opt}^{H})$, s must be equal to 0.25 while t becomes:

$$t = \frac{d_0 + d_1 k + d_2 k^2 + d_3 D_{dB} + d_4 D_{dB}^2 + d_5 D_{dB}^3}{1 + d_6 k + d_7 D_{dB} + d_8 D_{dB}^3} \tag{5-50}$$

Directivity is calculated from [9]

$$D_{dB} = f_0(k) + f_1(k) \ln(HP_E) \tag{5-51}$$

where

$$f_0(k) = f_{00} + f_{01} \ln(k) \tag{5-52}$$

and

$$f_1(k) = f_{10} + f_{11}\frac{1}{k} + f_{12}\frac{1}{k^2} + f_{13}\frac{1}{k^3} + f_{14}\frac{1}{k^4} + f^{15}\frac{1}{k^5} \tag{5-53}$$

D_{dB} and t are calculated from (5-51) and (5-50) respectively. From the parameters t, s and the HPBW in the two principal planes, A/λ and B/λ, from (5-44) and (5-45), are found. Finally, the axial length R is calculated from (5-40).

It is noticed that as in the previous case, the calculated results are approximate. The same scheme described in the previous section may be applied to find corrected values for A/λ. The iteration now takes place between Equations (5-39) and (5-44) for A/λ and t.

For $\min(k_{opt}^{E}, k_{opt}^{H}) \leq k \leq \max(k_{opt}^{E}, k_{opt}^{H})$ both of the above techniques have to be applied. One of the techniques gives the best solution.

Another design case is that where the ratio $k = \dfrac{HP_H}{HP_E}$ and the directivity of the horn are desirable. The design starts with the calculation of k_{opt}, which is the ratio of the HPBWs

Table 5.7 Coefficients of Equation (5-50)

d_0	d_1	d_2	d_3	d_4
0.1020	$2.9658.10^{-2}$	$-2.4894.10^{-3}$	$-2.0962.10^{-2}$	$6.3027.10^{-4}$
d_5	d_6	d_7	d_8	
$-5.9327.10^{-6}$	-0.6802	$-4.4039.10^{-2}$	$1.0213.10^{-3}$	

Table 5.8 Coefficients of Equations (5-52) and (5-53)

f_{00}	f_{01}	f_{10}	f_{11}	f_{12}	f_{13}	f_{14}	f_{15}
44.3672	-4.0980	-8.0775	-4.2683	14.5647	-26.1244	23.9791	-8.7301

of the unconstrained optimum horn to the desired directivity. The choice of the optimum parameter s or t and the use of the appropriate set of equations depend on whether k is bigger or smaller than k_{opt}. From Equation(5-47) or (5-51) and the desired k, the angles HP_H or HP_E are calculated. Knowledge of HP_H or HP_E and ratio $k = \dfrac{HP_H}{HP_E}$ give the desired HPBWs on both principal planes. The rest of the methodology described in the previous section can be followed.

5.3.5.3 Corrugated Pyramidal Horns

Corrugations on the walls of a pyramidal horn have been found to help in aperture tapering by reducing diffraction from horn edges and by lowering side lobe level. Several corrugations per wavelength with a depth of about $\lambda/4$ are necessary to achieve the above properties. The mode on the horn aperture is hybrid. A converter section is used to convert the waveguide mode into the hybrid one.

As in the smooth-walled case, the aperture field may be linearly or circularly polarized. In each of these cases, it can be expressed as

$$E_S = \hat{y} E_0 \cos \frac{\pi x}{A} \cos \frac{\pi y}{B} e^{-j\frac{\beta}{2}\left(\frac{x^2}{R_1}+\frac{y^2}{R_2}\right)} \tag{5-54}$$

for linear polarization and as

$$E_S = E_0 \cos \frac{\pi x}{A} \cos \frac{\pi y}{B} e^{-j\frac{\beta}{2}\left(\frac{x^2}{R_1}+\frac{y^2}{R_2}\right)} (K_{ry}\hat{y} + j K_{rx}\hat{x}) \tag{5-55}$$

for circular polarization.

K_{rx} and K_{ry} are factors that account for the different propagation constants of the hybrid modes:

$$K_{ry} = \sqrt{\beta^2 - \left(\frac{\pi}{A}\right)^2} \tag{5-56}$$

$$K_{rx} = \sqrt{\beta^2 - \left(\frac{\pi}{B}\right)^2} \tag{5-57}$$

For linear polarization, it is

$$P = P\hat{y} \tag{5-58}$$

Also, for circular polarization, it is

$$P = (j K_{rx}\hat{x} + K_{ry}\hat{y})P \tag{5-59}$$

P is given in the following form:

$$P = E_0 I_{1x} I_{2y} \tag{5-60}$$

I_{1x} and I_{2y} can be expressed with the help of Fresnel cosine and sine integrals as in the case of a smooth-walled horn:

$$I_{1x} = \frac{1}{2}\sqrt{\frac{R_1\pi}{\beta}}\left[e^{j\frac{\beta R_1}{2}(u+(\pi/\beta A))^2}[F^*(s_2)-F^*(s_1)] + e^{j\frac{\beta R_1}{2}(u-(\pi/\beta A))^2}[F^*(t_2)-F^*(t_1)]\right]$$

(5-61)

$$I_{2y} = \frac{1}{2}\sqrt{\frac{R_2\pi}{\beta}}\left[e^{j\frac{\beta R_2}{2}(v+(\pi/\beta B))^2}[F^*(s_2')-F^*(s_1')] + e^{j\frac{\beta R_2}{2}(v-(\pi/\beta B))^2}[F^*(t_2')-F^*(t_1')]\right]$$

(5-62)

where

$$\left. \begin{aligned}
t_2 &= \sqrt{\beta/\pi R_1}\left(A/2 - R_1 u + \frac{\pi R_1}{\beta A}\right)\\
t_1 &= \sqrt{\beta/\pi R_1}\left(-A/2 - R_1 u + \frac{\pi R_1}{\beta A}\right)\\
s_2 &= \sqrt{\beta/\pi R_1}\left(A/2 - R_1 u - \frac{\pi R_1}{\beta A}\right)\\
s_1 &= \sqrt{\beta/\pi R_1}\left(-A/2 - R_1 u - \frac{\pi R_1}{\beta A}\right)\\
t_2' &= \sqrt{\beta/\pi R_2}\left(B/2 - R_2 v + \frac{\pi R_2}{\beta B}\right)\\
t_1' &= \sqrt{\beta/\pi R_2}\left(-B/2 - R_2 v + \frac{\pi R_2}{\beta B}\right)\\
s_2' &= \sqrt{\beta/\pi R_2}\left(B/2 - R_2 v - \frac{\pi R_2}{\beta B}\right)\\
s_1' &= \sqrt{\beta/\pi R_2}\left(-B/2 - R_2 v - \frac{\pi R_2}{\beta B}\right)
\end{aligned} \right\}$$

(5-63)

Corrugated horns can be E- or H-sectoral. In this case, I_{1x} or I_{2y} are simpler. More specifically, for an E-sectoral horn, it is

$$I_{1x} = \frac{2A}{\pi\left(1-\left(\frac{Au}{\lambda}\right)^2\right)}\cos\left(\frac{\pi Au}{\lambda}\right)$$

(5-64)

I_{2y} is given in (5-62).

For an H-sectoral horn, it is

$$I_{2y} = \frac{2B}{\pi\left(1-\left(\frac{Bv}{\lambda}\right)^2\right)}\cos\left(\frac{\pi Bv}{\lambda}\right)$$

(5-65)

I_{1x} is given in Equation (5-61).

5.3.5.5 Setting Horns at ORAMA

By selecting 'Horn' as the element type, ORAMA derives the waveguides that operate in the fundamental mode at the 'Design Frequency'. The program displays waveguides

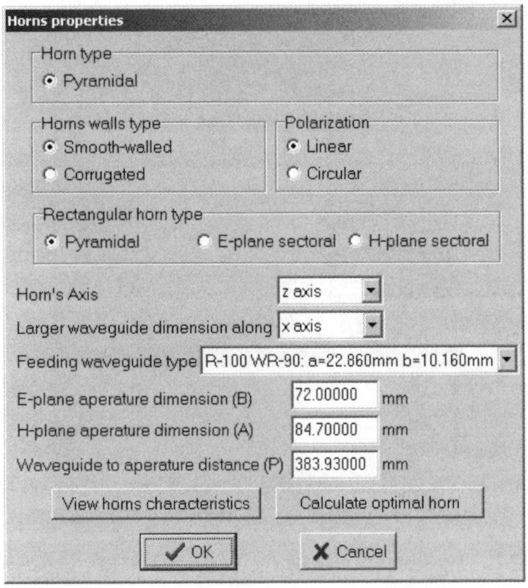

Figure 5.23 The pyramidal horn dialog box

in a drop-down list box entitled 'Feeding Waveguide Type' (see Fig. 5.23). If the design frequency is not covered in the range of international standard waveguides (EIA and IEC official types), ORAMA prompts the user to insert custom waveguide dimensions.

The items 'Horn walls type' and 'Polarization' allow selecting the type of horn walls (smooth-walled or corrugated) and the polarization (linear or circular) respectively.

Using the item 'Pyramidal horn type', one can choose a 'Pyramidal', an 'E-plane sectoral', or an 'H-plane sectoral' horn (see Fig. 5.23).

The next two drop-down list boxes determine the 'Horn's Axis' and the direction of the 'Larger waveguide dimension'. In the 'Feeding waveguide type' item, one can choose one of the available standard waveguides through the drop-down listbox. If the design frequency is outside the one covered by the available standard waveguides, ORAMA asks the user to enter the dimensions of a custom waveguide. The button 'Calculate optimal horn' derives the horn dimensions for three different choices. By pushing the above button, a dialog box appears (see Fig. 5.24).

In the dialog box, one can select to design the horn for a desired gain [10], or for certain values of HPBWs on the E- and H-plane, or for a desired gain and certain ratio of the HPBWs on the E- and H-plane. Depending on the choice, the dimensions of the horn are given in the 'Results found'. By pushing the button 'OK', one comes back to the dialog box of Fig. 5.23. The button 'View horns characteristics' moves the user to a dialog box where the pyramidal horn characteristics are given (see Fig. 5.25).

Going back to the main menu of ORAMA, one can calculate the pattern characteristics of a single element. If the characteristics are a little different from the predefined ones, iteration can be applied to change the values.

By using the smooth-walled horns characteristics as initial values and iterating the corresponding values of the corrugated horns, one can define their characteristics.

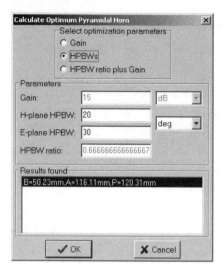

Figure 5.24   The dialog box of the optimum pyramidal horn design

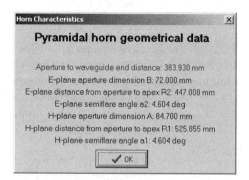

Figure 5.25   Pyramidal horn characteristics

Let us suppose that we have a desired frequency of 10 GHz. We choose the pyramidal horn as element type and we give the following design examples:

1. *Pyramidal horn with gain equal to 18 dB*: From ORAMA, we use the rectangular WR-90 with $a = 22.86$ mm and $b = 10.16$ mm as a feeding waveguide. The horn dimensions are found to be: $A = 101.32$ mm, $B = 78.055$ mm and $P = 94.624$ mm. By plotting the pattern of the designed horn, it is found that the resulting gain is 17.9 dB. To have the desired value of 18 dB, we increase the gain in the 'optimization parameters' to 18.1 dB. The new horn dimensions are: $A = 102.53$ mm, $B = 79.048$ mm and $P = 97.073$ mm. The resulting patterns of the horn are given in Fig. 5.26.

   If we choose the same dimensions for a corrugated horn, the resulting gain becomes equal to 17.42 dB. We design the horn again for 18.66 dB and the resulting gain for the corrugated horn becomes 18.0 dB. The horn dimensions are: $A = 109.57$ mm,

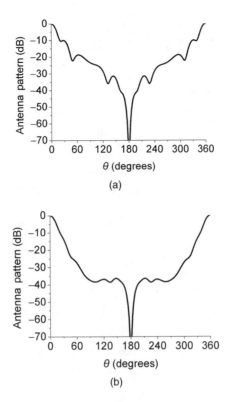

Figure 5.26 Patterns on (a) E-plane and (b) H-plane of a smooth-walled pyramidal horn with 18 dB gain

$B = 84.83$ mm and $P = 111.98$ mm. In Fig. 5.27, the patterns on the E- and H-plane are given.

2. *Pyramidal horn with desired HPBWS: $HP_H = 22°$ and $HP_E = 25°$* From ORAMA, we use again the rectangular WR-90 with $a = 22.86$ mm and $b = 10.16$ mm as a feeding waveguide. The horn dimensions are found to be: $A = 105.42$ mm, $B = 61.229$ mm and $P = 96.699$ mm. By plotting the pattern of the designed horn, it is found that the resulting HPBWs are: $HP_H = 21.997° \cong 22°$ and $HP_E = 25.035° \cong 25°$ (see Fig. 5.28).

   If we choose the same dimensions for a corrugated horn, the resulting HPBWs become $HP_H = 21.997° \cong 22°$ and $HP_E = 33.33°$. We design the smooth-walled horn again for $HP_E = 18.95°$ and the resulting HPBWs for the corrugated horn become $HP_H = 22°$ and $HP_E = 25°$. The horn dimensions are: $A = 105.42$ mm, $B = 86.95$ mm and $P = 96.70$ mm. In Fig. 5.29, the patterns on the E-and H-plane are given.

3. *Pyramidal horn with desired HPBW ratio equal to 1.2 and gain equal to 17 dB:* From ORAMA, we again use the rectangular WR-90 with $a = 22.86$ mm and $b = 10.16$ mm as a feeding waveguide. After iteration, for a gain of 17.14 dB, the horn dimensions are found to be: $A = 81.66$ mm, $B = 73.34$ mm and $P = 77.22$ mm. By plotting the pattern of the designed horn, it is found that the resulting gain is 17 dB and the

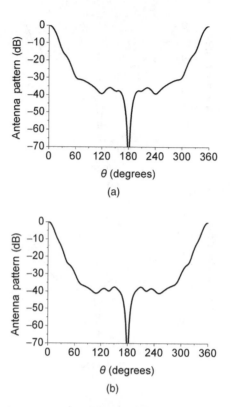

Figure 5.27   Patterns on (a) E-plane and (b) H-plane of a corrugated pyramidal horn with 18 dB gain

HPBWs are: $HP_H = 21.89°$ and $HP_E = 26.265°$ (see Fig. 5.30), which give a ratio equal ~ 1.2.

If we choose the same dimensions for a corrugated horn, the resulting gain becomes equal to 16.44 dB and $HP_E/HP_H = 1.1$. We design the smooth-walled horn again for a gain equal to 17.68 dB and $HP_E/HP_H = 1.1$. The resulting gain becomes 17 dB and the HPBW ratio for the corrugated horn becomes equal to 1.2. The horn dimensions are: $A = 100.67$ mm, $B = 76.36$ mm and $P = 87.04$ mm. In Fig. 5.31, the patterns on the E-and H-plane are given.

5.3.6 Microstrip Patches

In their simple form, microstrip patches consist of a metallic patch separated by a dielectric substrate from the ground plane [2]. The geometry of a rectangular patch is presented in Fig. 5.32.

The radiation of the patch comes from the fringing effect that occurs at the open transmission line ends (see Fig. 5.33).

If the thickness h of the substrate is $h \ll \lambda$, then the patch can be simulated by two radiating slots. Each slot has a uniform excitation with a length equal to the patch width (W). It also has a width equal to the thickness h of the substrate. Assuming that both

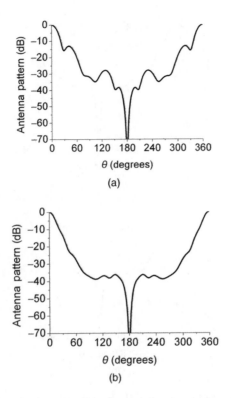

Figure 5.28 Patterns on (a) E-plane and (b) H-plane of a smooth-walled pyramidal horn with $HP_H = 22°$ and $HP_E = 25°$

slots are positioned on the xz plane, the radiation pattern is expressed by [2],

$$EF(\theta, \phi) = \sin\theta \frac{\sin\left(\frac{\beta h}{2}\cos\phi\sin\theta\right)}{\frac{\beta h}{2}\cos\phi\sin\theta} \frac{\sin\left(\frac{\beta W}{2}\cos\theta\right)}{\frac{\beta W}{2}\cos\theta} \cos\left(\frac{\beta L}{2}\sin\phi\sin\theta\right) \quad (5\text{-}66)$$

For a circular patch, the cavity model is used to derive the field at the edges. The radiation pattern is found by using the standard aperture theory [2]. It is

$$EF(\theta, \phi) = \sqrt{|E_\theta(\theta, \phi)|^2 + |E_\phi(\theta, \phi)|^2} \quad (5\text{-}67)$$

For the circular patch given in Fig. 5.34, the expression of the above quantities is

$$E_\theta(\theta, \phi) = [J_0(\beta a \sin\theta) - J_2(\beta a \sin\theta)]\cos\phi F_\theta(\theta, \phi) \quad (5\text{-}68)$$

$$E_\phi(\theta, \phi) = [J_0(\beta a \sin\theta) + J_2(\beta a \sin\theta)]\sin\phi \cos\theta F_\phi(\theta, \phi) \quad (5\text{-}69)$$

a is the patch radius and $F_\theta(\theta, \phi)$, $F_\phi(\theta, \phi)$ are correction factors due to the presence of the dielectric substrate[11].

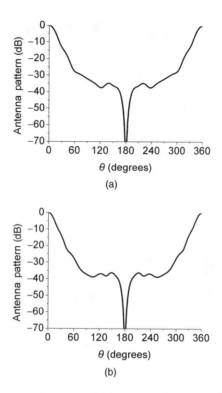

Figure 5.29 Patterns on (a) E-plane and (b) H-plane of a corrugated pyramidal horn with $HP_H = 22°$ and $HP_E = 25°$

5.3.6.1 Setting Microstrip Patches at ORAMA

Selecting 'Microstrip Patch' as the 'Element Type', one of the dialog boxes will appear, depending on the choice of rectangular or circular patch (see Figs. 5.35 and 5.36).

If a rectangular patch is selected (see Fig. 5.35), the width W, the length L and the thickness h of the patch must be specified. By using the two drop-down boxes, next to the dimensions of W and L, the orientation of the patch is specified. The dielectric constant of the substrate is not requested, because it is ignored in the calculations.

If a circular patch is selected (see Fig. 5.36), the radius a, and the thickness h of the patch must be specified. By using a drop-down box, the patch plane is specified. The dielectric constant of the substrate is specified in the last box.

5.4 Design Examples

Most of the examples presented in Chapter 4 have been obtained by using ORAMA. The reader can verify the examples and apply ORAMA to solve new problems. In this paragraph, some supplementary design cases will be given.

First, the pattern of a radar is presented. Suppose a cosecant pattern is desirable with maximum at 82° and a HPBW equal to ~13°. The SLL must be less than −25 dB. Figure 5.37 shows the patterns of the array for $N = 15$ rectangular patches ($W = 0.138\lambda$, $L = 0.15\lambda, h = 0.02\lambda$) and $d = 0.5\lambda$. The success of the OM is obvious.

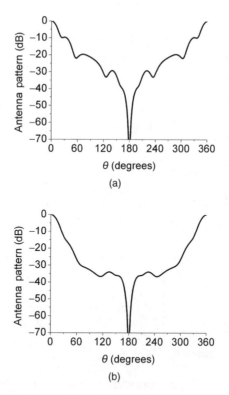

Figure 5.30 Patterns on (a) E-plane and (b) H-plane of a smooth-walled pyramidal horn with 17 dB gain and $HP_E/HP_H = 1.2$

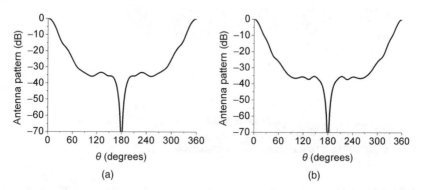

Figure 5.31 Patterns on (a) E-plane and (b) H-plane of a corrugated pyramidal horn with 17 dB gain and $HP_E/HP_H = 1.2$

The design case of a base station antenna is similar to the above given example. Suppose, it is desirable to illuminate an area with a base station antenna where the received signal at the mobile units is the same. It is known that a cosecant-square–shaped beam power pattern is necessary. Since there is a serious problem with the interference to other

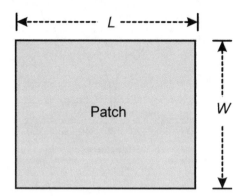

Figure 5.32   Geometry of a rectangular patch

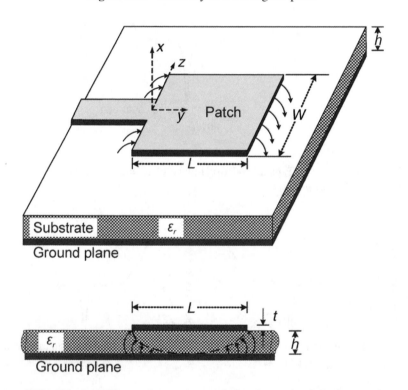

Figure 5.33   Fringing effect at the open transmission line ends of the rectangular patch

cells, a beam tilting procedure is used (see also in [12]). A six vertical collinear dipoles array with distance $d/\lambda = 0.86$ will be used. For a uniform excitation, the array has HPBW $= 9.89°$ and directivity $D = 10.14$ dBi $= 7.99$ dBd. For a tilting angle of $7°$ and a cosecant-squared–shaped beam power pattern, the excitation is given in Table 5.9. In the same table, the excitation of a similar case given in [13] is presented.

Figure 5.38 presents the pattern of the array.

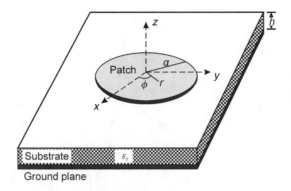

Figure 5.34   Geometry of a circular patch

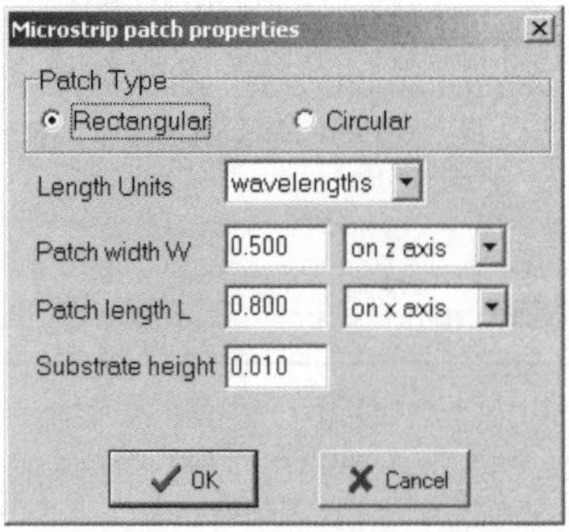

Figure 5.35   Dialog box for a rectangular patch

Figure 5.36   Dialog box for a circular patch

An 8-element linear array with $d/\lambda = 0.5$ is given next. A maximum at $\theta_0 = 70°$ is desired. In this case, the array factor can be written in the following form:

$$AF(\theta) = \delta(\theta - 70°) \tag{5-70}$$

The excitation of the elements of the array is uniform and the phase difference is $-61.56°$. This value is exactly the same as the corresponding one in Table 16.1 of [2].

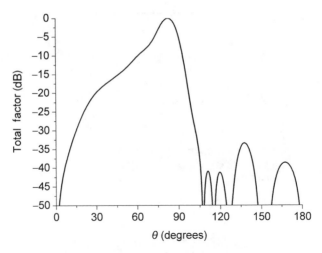

Figure 5.37   Cosecant pattern with maximum at $82°$ and HPBW $\sim 13°$ of a linear array of $N = 15$ rectangular patches and $d = 0.5\lambda$. The dimensions of each patch are: $W = 0.138\lambda$, $L = 0.15\lambda$, $h = 0.02\lambda$

Table 5.9   Normalized excitations for a six-dipole collinear array

Element number	[13]		[OM]	
	$\lvert A_i \rvert$	$\arg(A_i)$ (rad)	$\lvert A_i \rvert$	$\arg(A_i)$ (rad)
1	0.307	0.000	0.464	0.000
2	0.372	0.175	0.888	0.180
3	0.438	0.393	1.605	0.288
4	0.438	0.655	1.605	1.338
5	0.438	1.384	0.888	1.446
6	0.438	2.845	0.464	1.626

In [2], the angle θ_0 is written as $90° - 70° = 20°$. If it is also desirable to have a null at $\theta_1 = 45°$, then the null constraint is applied. In this case, the excitation is non-uniform and is found to be exactly the same as that given in Table 16.2 of [2]. The same array with maximum at $\theta_0 = 90°$ and null at $\theta_1 = 45°$ has a pattern presented in Fig. 5.39. This pattern is exactly the same as that of Fig 16.43 in [2] with the angle θ_0 at $0°$. If it is desirable to have a Chebyshev pattern with SLL $= -26$ dB and null at $\theta_1 = 45°$, then the pattern is also given in Fig. 5.39. This pattern shows a small increase in the SLL on the side where the constraint is applied.

    The design of a linear array with 15 rectangular horns at a distance $d/\lambda = 0.88$ is given next. The working frequency of the array is at $f = 6$ GHz. A rectangular waveguide WR-159 feeds the horn ($a = 40.39$ mm, $b = 20.19$ mm). The size of the rectangular horn for a directivity of 6 dB is $A = 44.21$ mmX$B = 24.741$ mm. It is desirable to have a Chebyshev broadside pattern with SLL < -20 dB. A $T_{14}(x)$ polynomial with the appropriate SLL is used. The pattern of the array is shown in Fig. 5.40. If it is desirable to rotate the beam by $4°$, then the new pattern is presented in Fig. 5.41. It is observed that a side lobe

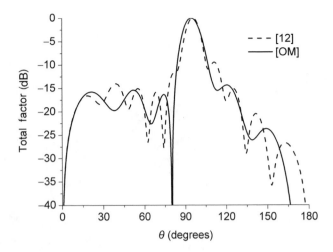

Figure 5.38 Cosecant-squared–shaped beam pattern of a six collinear dipoles array with tilting angle of $7°$ and $d/\lambda = 0.86$

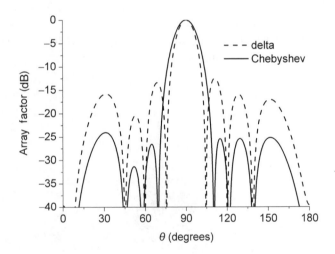

Figure 5.39 Radiation patterns of a linear array with $N = 8$ and $d/\lambda = 0.5$. The desired patterns are: A delta function and a Chebyshev polynomial with maximum at $90°$ and null constraint at $45°$

at $180°$ appears. A beam rotation for more than $5°$ produces an SLL more than -20 dB, which is out of the design assumptions.

Two design cases of a 32-element linear array with $d/\lambda = 0.5$ will be given next. We start from a uniform broadside array where there are two areas of minimization at $[0° - 80°]$ and $[100° - 180°]$. The resulting pattern is represented in Fig. 5.42. Comparison of the above figure with that given in Fig. 16.5 of [14] shows lower-side lobes.

The same array with null constraints at $150°$, $160°$ and areas of minimization at $[20° - 30°]$ and $[120° - 140°]$ gives a pattern represented in Fig. 5.43. It is obvious that the array follows the constraints.

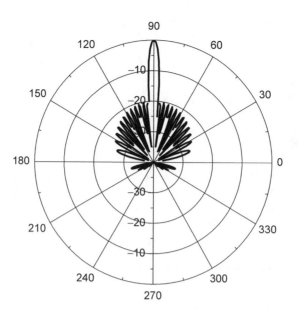

Figure 5.40   Chebyshev pattern with maximum at $\theta = 90°$ of a linear array with 15 rectangular horns at a distance $d/\lambda = 0.88$. The working frequency is at $f = 6$ GHz. The rectangular horn for a directivity of 6 dB has $A = 44.21$ mm and $B = 24.741$ mm

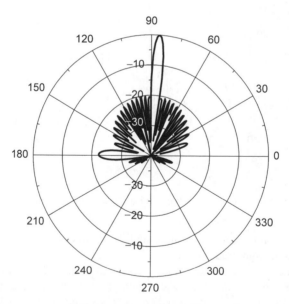

Figure 5.41   Chebyshev pattern with maximum at $\theta = 86°$ of the array given in Fig. 5.48

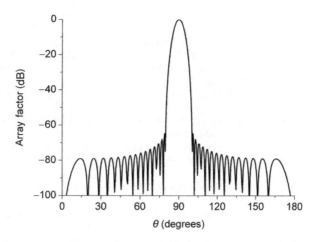

Figure 5.42 Pattern of a linear array with 32 elements in $d/\lambda = 0.5$ with two areas of minimization at $[0° - 80°]$ and $[100° - 180°]$

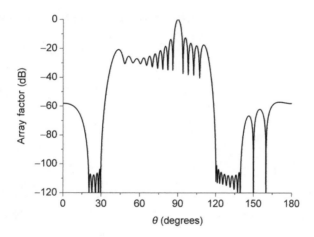

Figure 5.43 Pattern of a linear array with 32 elements in $d/\lambda = 0.5$ with null constraints at 150°, 160° and areas of minimization at $[20° - 30°]$ and $[120° - 140°]$

5.5 Conclusion

The ORAMA tool is a computer program that can be used for various design cases as it is useful for wireless communications, radar and broadcasting. The CD that is attached to this book contains one of the versions of ORAMA designed in our Radio-communications Laboratory (RCL).

It is highly recommended that, after a little practice, the reader should try to solve his own problems by using ORAMA.

Please note that incompatible selections of the number of elements and the geometry of the array results in unsuccessful patterns. Large number of elements and/or small mutual distance between them may produce inaccurate results.

References

[1] J.N. Sahalos, *Antennas*, Aibazis-Zoumpoulis, Thessaloniki (in Greek), 1989.

[2] C.A. Balanis, *Antenna Theory, Analysis and Design*, 3rd ed., John Wiley & Sons, New York, 2005.

[3] J.N. Sahalos, "On the Design of Antennas Consisting of Arbitrarily Oriented Dipoles," *IEEE Trans. Antennas Propag.*, Vol. **AP-24**, pp. 322–327, 1976.

[4] M. Calvo, "Primary Feeds for Reflectors and Lenses," in *Reflector and Lens Antennas*, Artech House, Norwood, MA, 1988.

[5] A.W. Lowe, A.W. Rudge and A.D. Olver, "Primary Feed Antennas," in *The Handbook of Antenna Design*, Vol. I, Chap. 4, Peter Peregrinus, London, 1982.

[6] J.N. Sahalos, *Microwaves*, Aibazis-Zoumpoulis, Thessaloniki, (in Greek), 1990.

[7] M. Abramowitz and I.A. Stegun (Eds.), *Handbook of Mathematical Functions*, Dover Publications Inc, New York, 1972.

[8] W.H. Press, B.P. Flannery, S.A. Teukolsky and W.T. Vetterling, *Numerical Recipes in C*, Cambridge University Press, MA, 1988.

[9] G. Kordas, K.B. Baltzis, G.S. Miaris and J.N. Sahalos, "Pyramidal Horn Design Under Constraints on Half-Power Beamwidth," *IEEE Antennas Propag. Mag.*, Vol. **44**, pp. 102–108, 2002.

[10] T.A. Milligan, *Modern Antenna Design*, McGraw-Hill, New York, 1985.

[11] R.A. Saitani, *CAD of Microstrip Antennas for Wireless Applications*, Artech House, Norwood, MA, 1994.

[12] K. Fujimoto and J.R. James, *Mobile Antenna Systems Handbook*, Artech House, Norwood, MA, 1994.

[13] K. Siwiak, *Radiowave Propagation and Antennas for Personal Communications*, Artech House, Norwood, MA, 1995.

[14] L.C. Godara, *Handbook of Antennas in Wireless Communications*, CRC Press, New York, 2002.

Index

CHECK FOR _____|_____ CDS.